Advances in Oil and Gas Exploration & Production

Series Editor

Rudy Swennen, Department of Earth and Environmental Sciences, K.U. Leuven, Heverlee, Belgium

The book series Advances in Oil and Gas Exploration & Production publishes scientific monographs on a broad range of topics concerning geophysical and geological research on conventional and unconventional oil and gas systems, and approaching those topics from both an exploration and a production standpoint. The series is intended to form a diverse library of reference works by describing the current state of research on selected themes, such as certain techniques used in the petroleum geoscience business or regional aspects. All books in the series are written and edited by leading experts actively engaged in the respective field.

The Advances in Oil and Gas Exploration & Production series includes both single and multi-authored books, as well as edited volumes. The Series Editor, Dr. Rudy Swennen (KU Leuven, Belgium), is currently accepting proposals and a proposal form can be obtained from our representative at Springer, Dr. Alexis Vizcaino (Alexis.Vizcaino@springer.com).

Timothy Tylor-Jones · Leonardo Azevedo

A Practical Guide to Seismic Reservoir Characterization

Timothy Tylor-Jones
Production and Operations
bp
Sunbury, UK

Leonardo Azevedo
CERENA/DECivil
Instituto Superior Técnico
Universidade de Lisboa
Lisboa, Portugal

ISSN 2509-372X   ISSN 2509-3738   (electronic)
Advances in Oil and Gas Exploration & Production
ISBN 978-3-030-99856-1   ISBN 978-3-030-99854-7   (eBook)
https://doi.org/10.1007/978-3-030-99854-7

We dedicate this book to our families

Acknowledgements

The original concept of this book was developed after a joint research project conceptualized and developed under the supervision of both authors. At that time, the main idea was to write a reference book that could be used by graduate students and young professionals joining the energy industry. The text would simultaneously provide a practical perspective and enough theoretical background on modern geo-modelling workflows focusing on the steps related to subsurface modelling and characterization using seismic reflection data. Hopefully, we were able to accomplish such objectives by leveraging our experience within the energy industry and academia. The development of each chapter reflects our complementary backgrounds with all the pros and cons. This book is therefore the result of experience acquired during our professional paths and consequently there are several people that need to be acknowledged. The first should be Tiago Ribeiro who first introduced both of us and then Mariana Martinho who succeeded in getting an MSc. degree under our supervision.

The bringing together of so many different concepts of geophysics required a broad and deep knowledge of both theory and practical application. I would like to acknowledge the guidance, support and mentoring of John Etgen, Mel Ball, Walter Rietveld, Sarah Buchanan who took time out of their lives to help the authors with this endeavour. A special mention must be made to Felicity Tylor-Jones whose exceptional skills as a professional copy editor and proof reader have transformed the reading experience of such a technical and at times complex subject into something very readable.

An important contribution to this book originated from the research and development performed at the Centro de Recursos Naturais e Ambiente (CERENA) from Instituto Superior Técnico (Lisboa, Portugal) on topics related to geostatistical modelling of the subsurface. The methods introduced in Chaps. 7 and 8 are an extension of the pioneering work of Prof. Amílcar Soares in coupling geophysics and geostatisics for subsurface modelling and characterization. Amílcar and Prof. Maria João Pereira have been mentors and friends of Leonardo Azevedo during the last decade. Without their contributions and continued support this book would not be possible. LA also thanks all the Ph.D. and MSc. students who he was lucky to meet and work with during these years. Some of the ideas expressed in this book are directly related to their research work. These students have benefited from the industry sponsors of the group, to which LA acknowledges. A special acknowledge goes to João Narciso and Roberto Miele for their detailed revision of the manuscript and insightful comments.

Finally, thanks to our families for the support and encouragement during the writing of this book.

Timothy Tylor-Jones
Leonardo Azevedo

Contents

Acronyms

AGC	Automatic gain control
AIGI	Acoustic impedance, gradient impedance
AVO	Amplitude versus offset
AVA	Amplitude versus angle
CCUS	Carbon capture, usage and storage
CDP	Common depth point
CMP	Common-midpoint
Co-DSS	Direct sequential co-simulation
CPI	Computer Processing Interpretation
dB	Decibel
DDI	Direct detection indicator
DHI	Direct hydrocarbon indicator
DSS	Direct sequential simulation
EEI	Extended elastic inversion
EGS	Engineered geothermal system
EI	Elastic impedance
EM	Expectation Maximization
FWI	Full Waveform Inversion
GDE	Gross depositional environment
GPS	Global positioning system
HCI	Hydrocarbon indicators
HPTP	Highpressure high temperature
LFM	Low-frequency model
LNR	Linear noise removal
LSQ	Least Square migration
LWD	Logging While Drilling
MEMS	Micro-electro-mechanical system
MAP	Maximum a posteriori model
McMC	Markov chain Monte Carlo
NMO	Normal moveout
NFH	Near-field hydrophone
OBN	Ocean bottom nodes
OBS	Ocean bottom seismic
OCB	Ocean bottom cable
OWT	One-way travel time
PBR	Primary-to-bubble ratio

PEP	Percentage of energy predicted
QC	Quality control
RC	Reflection coefficient
RMO	Residual normal moveout
RMS	Root-mean-square
ROV	Remotely operated vehicle
RTM	Reverse time migration
SCAL	Special core analysis
SEG	Society of Exploration Geophysicists
SEM	Scanning electronic microscopy
SGS	Sequential Gaussian simulation
TWT	Two-way travel time
VSP	Vertical seismic profile
VTI	Vertical transverse isotropic
WAZ	Wide azimuth survey
I_P	P-impedance
I_S	S-impedance
V_P	P-wave propagation velocity
V_S	S-wave propagation velocity
V_{sh}	Volume of shale
S_w	Water saturation
ϕ	Porosity
ρ	Density

List of Figures

Introduction

1

The quest to understand the internal structure of the Earth goes back to ancient times, because understanding how the world we live in works is part of human nature. Geophysics, as the science that studies the physical processes and properties of the Earth, contributes to this objective. It is believed that the first seismological detector was built in A.D. 132 by the Chinese philosopher Chang Heng. Through time, geophysical principles were widely applied in several areas and played a central role in navigation. However, only in the nineteenth century was geophysics recognized as an individual discipline, and in the following century was broadly applied to explore the solid and liquid Earth.

Today, exploration geophysics is key to understanding Earth processes at different spatial scales, from the near surface down to the interface between the crust and the mantle and the internal structure of the planet earth. Geophysics has been widely applied to the exploration of natural resources such as oil and gas reservoirs, mineral deposits and the characterization of groundwater, soil and contaminated sites. Exploration geophysics is not limited to these rather traditional areas but has the potential to be a critical asset towards carbon neutrality (e.g., for gas storage).

To this end, the seismic reflection method (i.e., the study of the interior of the Earth with seismic waves, which are reflected at geological boundaries with different elastic properties) has been the preferred geophysical technique to investigate the subsurface in detail. Seismic reflection has proven its value for the modelling and characterization of hydrocarbon accumulations and in mineral deposits. However, the value of seismic reflection data is not limited to these fields. This type of data is already mandatory for geological carbon storage studies to ensure economic viability of projects and the long-term safety of potential storage sites; for the geotechnical characterization of the seafloor sediments for offshore windfarms installation; and for the mapping of subsurface units to support the viability and drilling of boreholes to access geothermal energy.

The value of seismic reflection data is undeniable when creating numerical three-dimensional models of the subsurface and particularly for the prediction of the spatial distribution of the subsurface elastic (i.e., density, P-wave and S-wave velocities) and rock properties (e.g., fluid saturation, porosity and mineral fractions) in locations where direct observations (i.e., borehole data) are not available. In other words, these data play a major role in the so-called geo-modelling workflow when interpreted from a quantitative perspective. This is often designated in the literature as quantitative seismic interpretation or reservoir geophysics. However, data processing and integration with geological modelling tools is not straightforward, as the seismic reflection data are indirect measurements of the true subsurface geology. Its usage requires experience in a set of advanced technological tools.

T. Tylor-Jones and L. Azevedo, *A Practical Guide to Seismic Reservoir Characterization*, Advances in Oil and Gas Exploration & Production, https://doi.org/10.1007/978-3-030-99854-7_1

With this book we aim at two main objectives: (i) to provide an introductory level for graduate students and/or early career geophysicists by combining practical aspects with the most advanced techniques for quantitative seismic interpretation; (ii) to provide a description of the most common techniques applied from data pre-conditioning to modelling (i.e., seismic inversion).

In the next chapters we review the basics about the seismic method, including acquisition and processing, and we set the stage for the data pre-conditioning steps required to use these data in the geo-modelling workflow. This is followed by a description of the most common wavelet extraction methods and their advantages and disadvantages. In Chap. 6 we provide the background for amplitude versus offset (AVO) analysis and the interpretation of seismic amplitudes. In Chaps. 7 and 8 we introduce advanced methods on how seismic data can be used to predict the spatial distribution of the subsurface elastic and rock properties. We finish with our current perspective of what the future of these

data might be and their role in the current energy transition towards carbon neutrality.

Before we start, we provide a general and practical workflow that overviews all the steps described in detail in the following chapters.

1.1   The General Workflow for Seismic Reservoir Characterization

Figure 1.1 is a schematic representation of a conceptual workflow followed by a geoscientist to predict the spatial distribution of the subsurface rock properties for energy resources characterization (e.g., hydrocarbon and CO_2 reservoir modelling). This is a multidimensional workflow which comprises multiple steps that might exist, or not, depending on the data availability and their quality (Pyrcz and Deutsch 2002). The steps described in detail in this book are those represented in light green in Fig. 1.1. They comprise a rather classic interpretation of the data for mapping structures (Brown 2011) that are more likely

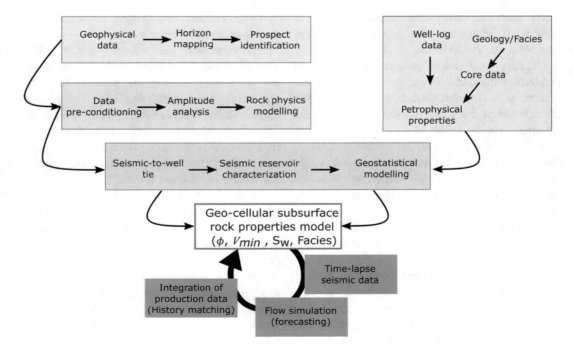

Fig. 1.1 A general workflow for subsurface modelling and characterization

to contain structures able to accumulate and store fluids. This initial stage finishes in the prospect characterization and de-risking.

Next, we move forward aiming to extract geological information from the reflected seismic amplitudes. At this stage, we might need to revise the seismic data processing applied to the data, and then study and model the variation of reflection amplitudes with the incident angle and model their change due to geological factors such as the geopressure gradient and compaction trends. The integration of direct observation (e.g., borehole data) of the subsurface is critical as these data provide valuable information not captured by seismic data. While this book does not deal with the processing and interpretation of these data, they are important inputs for the quantitative seismic interpretation step. The ultimate goal is to model the spatial distribution of the most fluid-relevant rock properties within a geo-cellular model of the subsurface (i.e., a numerical three-dimensional model of the sub-surface geology).

The resulting model, or models, are later used to forecast the behaviour of the system under investigation after producing and/or injecting fluids. After drilling, fluid production or injection starts, these additional data are a valuable window of information about the subsurface geology and are used to tune and update the model (i.e., history matching) (Oliver et al. 2008). These data can be coupled with time-lapse seismic reflection data to better predict the fluid behaviour in the subsurface.

References

Brown AR (2011) Interpretation of three-dimensional seismic data. Society of Exploration Geophysicists and the American Association of Petroleum Geologists

Oliver DS, Reynolds AC, Liu N (2008) Inverse theory for petroleum reservoir characterization and history matching. Cambridge University Press

Pyrcz MJ, Deutsch C (2002) Geostatistical reservoir modeling, 2nd edn. Oxford University Press, New York

The Seismic Method

2.1 Basics of Wavefield Propagation

Geophysicists or Earth scientists working in the field of seismology are concerned with the detection of seismic waves (e.g., sound waves) using a seismometer. These waves originate from a single source and propagate away as energy disturbances in the form of waves through the surrounding elastic medium (e.g., water, air or the rock). In other words, seismic waves are travelling vibrations or oscillations caused by the displacement of particles of the medium through which they move.

To record and study seismic wave data a seismic event must occur to generate energy, which can be created from natural or artificial (i.e., man-made) sources. When seismic waves are generated from natural sources, such as the release of built-up energy via an earthquake, large-scale rockfalls or even a volcanic eruption, the amount of energy released can be huge; much larger than anything required for use in commercial seismic surveys within the energy industry. However, large natural seismic events have been the mechanism for many advances in seismology over the centuries and have led to increases in our understanding about important scientific topics like the internal structure of our own planet. In fact, the origin of the word seismic comes from the ancient Greek *seisomós* meaning shaking earthquake.

Natural seismic events have great value in showing geoscientists about the structure and propagation velocity contrasts below the Earth's surface at both shallow and great depths. To put this in perspective, the average thickness of the Earth's crust varies from 5–10 km for oceanic crust up to 30–45 km for continental crust, yet represents only $\sim 1\%$ of the thickness of the Earth. The deepest mine on Earth is the Mponeng gold mine in South Africa with a depth of just 3.5 km, and the deepest borehole is the Kola Superdeep in Russia at 12.262 km deep (with a bottom hole diameter of 23 cm). These represent the current maximum technological limits and depths of investigation for physical exploration into the Earth's crust. Using seismic waves, we can explore indirectly far deeper—right into the centre of the Earth. Geophysical techniques allow us to understand the complex structures inside our own planet which we can never hope to physically sample due to the extreme pressures and temperatures.

Natural seismic events are both unpredictable and random, which makes them unreliable as a regular seismic source for repeated subsurface investigation. In addition, the magnitude of natural seismic wave properties are highly variable. For use in commercial applications, man-made seismic waves ensure the amount of energy released by an artificial seismic source is bespoke depending on the objective of the survey and the expected depth of the target.

© The Author(s), under exclusive license to Springer Nature Switzerland AG 2022
T. Tylor-Jones and L. Azevedo, *A Practical Guide to Seismic Reservoir Characterization*,
Advances in Oil and Gas Exploration & Production, https://doi.org/10.1007/978-3-030-99854-7_2

In the next sections we describe in detail the type of seismic waves used and generated artificially in a seismic reflection survey (i.e., exploration geophysics) and the physics that govern the behaviour of these waves. For introductory books on exploration geophysics, we refer to Dobrin (1976), Sheriff (1991), Sheriff and Geldart (1995), Kearey et al. (2002).

2.1.1 Stress and Strain

When seismic waves propagate, or travel, away from the point of origin (i.e., source) they must pass through the surrounding medium. As the energy waves come into contact with the subsurface physical particles (i.e., the grains of the rocks) a force is exerted onto them which alters the natural equilibrium of forces acting on the particles. Seismic waves subject rocks within the subsurface to stress and strain as a consequence of their propagation. We can understand stress as a magnitude of force being applied to an area of material which results in changes to the material, either temporary or permanent, that are measured as strain. Evidence that a material has experienced strain is usually in the form of deformation, and can be defined as the change of size or shape of an object compared to its original state (Fig. 2.1).

Stress science defines that at any point on a material under stress there are three orthogonal stress planes called principal axes of stress. Along these axes normal stress is applied in a perpendicular direction to a cross section of the material. Normal stress can be split into two types: tensile stress and compressional stress (Fig. 2.1a). Tensile stress acts in the tensile direction to stretch or lengthen the material. For compressional stress, the direction of stress compresses or shortens the material. The stress is compressional if the forces are directed towards each other (Fig. 2.1b) and tensile if they are directed away from each other. The strain resulting from normal stress can also be categorized with similar terminology: tensile strain and compressional strain. When a force acts parallel to a material surface, rather than perpendicular,

stress and strain are produced by shear forces. Shear stress still acts on the material surface, however the resulting strain is more a measure of maximum displacement because the material deformation is not uniform. Shear strain is measured as an angle that quantifies the maximum displacement away from the position prior to shear stress being applied (Fig. 2.1c).

When rock in the subsurface is subjected to stress caused by seismic waves this can be summarized by one stress equation and three main directional strain equations directly related to the type of strain:

$$\text{Normal Stress} = \frac{\text{force}}{\text{area}}, \qquad (2.1)$$

$$\text{Longitudinal strain} = \frac{\text{change in length}}{\text{original length}}, \quad (2.2)$$

$$\text{Volume or bulk strain} = \frac{\text{change in volume}}{\text{original volume}}, \\ (2.3)$$

$$\text{Shear strain} = \text{shear angle } (\theta), \text{in radians.} \\ (2.4)$$

2.1.2 Elastic Moduli

Hooke's[1] law is undoubtedly considered the underpinning of modern elastic wave theory. His work on the relationships between strain and stress, commonly referred to as Hooke's law, was first published in "Potentia Restitutiva" (1678) and states that the force (F) needed to extend or compress a spring by some distance (x) is linearly proportional to that distance:

$$F = kx, \qquad (2.5)$$

where k is defined as the stiffness of the spring and x is relatively small to the total possible deformation of the spring.

Hooke's work explains the motion of a spring as it moves up and down with a weight attached

---

[1] Robert Hooke born on the Isle of Wight in England in 1635 and died in 1703.

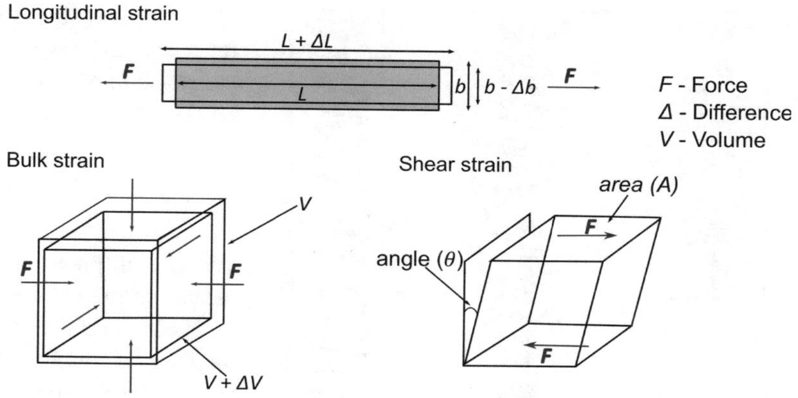

Fig. 2.1 Schematic representation of longitudinal, bulk and shear strain

to its end, which pertains to the relationship between stress and strain for a linear elastic solid (Fig. 2.2). Hooke observed that an elastic material always requires the same amount of force to produce the same amount of strain (i.e., stretching in length which is dependent on a material's stiffness). This means that the strain is directly proportional to the stress within an elastic material. Hooke's law allows us to define the elastic moduli that corresponds to the directional strain.

How a solid material behaves when placed under stress is not as simple as Hooke's law may first suggest. Initially, when a small force is applied the relationship between stress and strain can be described as linear (Fig. 2.3). However, as the stress increases the material will move into a state known as ductile, where the relationship between stress and strain is no longer linear and a small increase in stress results in a great change in strain. With increasing stress, at some point, the material will reach its fracture point where the material physically separates and represents the point of maximum strain. The ultimate strength point of the material is defined when the material weakens but does not necessarily break and can lead to a reduction of stress on the material prior to the fracture point (Fig. 2.3). The strain in seismic waves is of such a small magnitude that it falls within the elastic deformation zone (Fig. 2.3).

There are a number of standard elastic moduli used to define the effects of stress and strain on

materials and they are expressed as ratios of a specific stress type and resulting strain. Seismic waves induce minute amounts of strain on the rocks as they travel through the subsurface and the resulting deformation is elastic. These moduli are particularly relevant for subsurface characterization from seismic waves because the variability of seismic wave velocity travelling through the subsurface is related to both the density and the elastic properties of different rocks. Calculating elastic moduli therefore helps with identifying and comparing different elastic behaviours of rocks, which can be related to variations in lithology and pore-fluid types. The selection of elastic moduli used during interpretation should be made based on rock physics modelling and understanding of optimum sensitivity to the relevant lithologies and fluids.

Three fundamental properties are required to derive the majority of elastic moduli: P-wave velocity, S-wave velocity and density. The parameters of P-wave and S-wave velocities can be measured from both laboratory and field data (e.g., core samples, well-logs and seismic). Measuring density from core and well-logs is also straightforward, however estimating these properties from seismic data requires running a seismic inversion (Chap. 7) workflow, which presents some challenges associated with the reliability of the resulting product (Avseth et al. 2005). If the geoscientist has access to seismic inversion products in the form of P-impedance, S-impedance, P-velocity to S-velocity ratio and

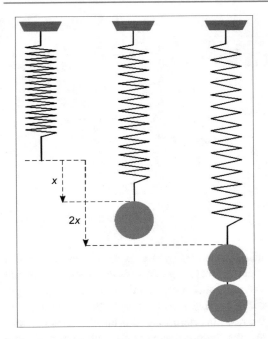

Fig. 2.2 Schematic representation of Hooke's law

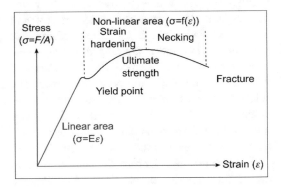

Fig. 2.3 Relationship between stress and strain

directions and surfaces. This stress results in the reduction of total volume of the material. Once the stress is removed, the material returns to its original volume. The bulk modulus is defined as:

$$K = \frac{P}{\Delta V / V}, \qquad (2.6)$$

where P is pressure, V the volume and ΔV a change in the volume.

Shear modulus (μ) is also known as the modulus of rigidity. The movement of shear is a sliding of the top and base of a solid. Shear modulus is of particular interest because of the inability to record shear modulus in liquids. When shear stress is applied to a liquid the resulting shear strain is not measurable because it continues to deform and does not reach a new strain equilibrium state for measurement; the shear modulus of liquid is, therefore, essentially zero. This phenomenon becomes especially important when looking at fluid properties of rocks with recorded S-waves because, unlike P-waves, they have only passed through the rock framework and matrix, not through the liquid-filled pore spaces. The ratio of P-wave to S-wave velocities (V_P/V_S) is commonly used to distinguish different pore fluids. S-waves are not always recorded in a seismic survey but they can be derived from P-wave velocities using empirical and theoretical methods for a particular rock medium. Shear velocity information for seismic reservoir characterisation is commonly combined with density to create S-impedance.

The shear modulus describes the ratio of shear stress to shear strain where the shear strain is measured by the shear angle:

$$\mu = \frac{\text{shear stress}}{\text{shear strain}}. \qquad (2.7)$$

Young's modulus[2] (E) is referred to as the modulus of elasticity and can be used to calculate the ratio of longitudinal stress to strain of an

density then calculating a range of different elastic moduli should be relatively straightforward. Geoscientists may prefer to interpret certain elastic moduli based on previous experience, while some elastic moduli have become synonymous with their tendency to be sensitive to particular properties (Goodway et al. 1997).

Bulk modulus (K) is the elastic modulus constant used to describe the elastic properties of a solid material which is subjected to stress in all

isotropic material. All materials have a measure of Young's modulus that vary depending on temperature and pressure and provides information about when a material will deform. It is therefore also useful to describe how the rocks in the subsurface deform elastically as seismic waves propagate through them, identifying differences in lithology and pore fluids. Young's modulus is also able to give information about how ductile or brittle subsurface rocks are. Understanding brittleness is of particular interest when looking to induce artificial fractures within a specific brittle formation, which are optimum for fracture stimulation (Chopra and Marfurt 2005).

Axial modulus, sometimes referred to as a special type of Young's modulus, is described as stress being applied along the longitudinal direction with minimal lateral stress, or, as a ratio of longitudinal stress to uniaxial longitudinal strain. This is like the propagation of a P-wave (Sect. 2.1.3):

$$E = \frac{F/A}{\Delta l/l},\qquad(2.8)$$

where F is the stretching force, A is the cross-sectional area and l is the original length.

Poisson's ratio[3] (σ) describes the strength of a material by giving a ratio of lateral strain against longitudinal strain. The simplest way to think about Poisson's ratio is pulling both ends of an elastic tube, but, unlike Young's modulus, the material gets thinner in cross section (i.e., lateral strain) the more it is pulled apart (i.e., longitudinal strain). Poisson's ratio can be defined as a function of the bulk modulus and elasticity showing the resulting strain with the following equation:

$$\sigma = \frac{3K - 2\mu}{2(3K + \mu)},\qquad(2.9)$$

where K is the bulk modulus and the μ is shear modulus.

However, rather than using the above equation for seismic reservoir characterization, the application of Poisson's ratio is better considered in terms of velocities using the ratio of V_P to V_S. Verm and Hilterman (1995) simplify the AVO Shuey equation (Table 6.2) to produce the attribute Poisson's reflectivity, which is concerned with change of Poisson's ratio for material layers above and below an interface. Quakenbush et al. (2006) identify that the incorporation of density estimation together with Poisson's ratio has additional benefits for reservoir delineation, and proposes a combined attribute called Poisson's impedance:

$$PR = \left[\frac{(\sigma_2 - \sigma_1)}{(1 - \sigma_{avg})^2} \right]\qquad(2.10)$$

where $\sigma_2 - \sigma_1$ are the Poisson's ratio above and below a given geological interface and the σ_{avg} is the average Poisson's ratio for the two layers above and below the interface.

When discussing both bulk and shear moduli, Lamé's[4] elastic constants should also be considered. Lamé's two elastic constants λ and ε are widely associated with a capacity to detect and differentiate lithology and fluid types when compared to methods using only compressional (P-wave) and shear (S-wave) wave velocities. Lamé's elastic constants can be used to focus more on changes in velocity rather than displacement and stress across an interface between different rock properties of the propagating seismic wave from the wave equation (Goodway et al. 1997). Goodway et al. (1997) came up with the Lambda-Mu-Rho method which uses crossplots of the properties Lambda-Rho against Mu-Rho to differentiate lithology and fluid types. The Lambda-Mu-Rho method has been further enhanced as a hydrocarbon fluid indicator with the inclusion of poro-elastic theory (Russell et al. 2003) which partitions the elastic response of the reservoir into a solid and a fluid component, respectively. This method has a definite physical

---

[3] Named after the French mathematician and physicist Simeon Poisson (1781–1840).

[4] Gabriel Lamé was a French mathematician (1795–1870) with an interest in the mathematical theory of elasticity, specifically looking at stress–strain relationships.

meaning, which reduces the ambiguity in fluid detection (Zhou et al. 2021):

$$\lambda = \frac{vE}{(1+v)(1-2v)},$$

$$\lambda = K - \frac{2}{3}G,$$

$$\lambda = \frac{2vG}{1-2v},$$

$$\lambda = 3K\frac{v}{1+v},$$

$$\lambda = \rho\left(V_p^2 - 2V_s^2\right) \qquad (2.11)$$

$$\mu = \frac{E}{2(1+v)},$$

$$\mu = \frac{3}{2}(K - \lambda),$$

$$\mu = \lambda\frac{1-2v}{2v},$$

$$\mu = 3K\frac{1-2v}{2+2v},$$

$$\mu = \rho V_s^2.$$

2.1.3 Types of Waves

Seismic wave types are classified based on their velocity and the geometry of particle motion they produce. Seismic waves can be broadly divided into body waves (i.e., P-waves and S-waves) and surface waves (i.e., Rayleigh waves and Love waves), but there are further variations such as guided, interface and transversal waves. When looking at man-made seismic waves, from surveys used for the application of finding natural resources, geoscientists will mostly deal with body waves. However, it is useful to have knowledge of the different wave types, particularly for seismic data processing as a variety of waves may be present in the recorded data that will require identification and removal.

Body waves that propagate through an elastic solid medium cause particles to oscillate in

periods of dilatation and compression about their resting equilibrium before passing on to the next particle and so on. This extension and compression can be interpreted as variations of particle velocity and displacement of space–time. Body waves in a seismic experiment can be subdivided into two types (Fig. 2.4).

Primary (P-waves), or compressional, waves are the fastest and first to arrive at a recording station from the source origin. The stress induced from this type of wave leads to a compressional strain, or push–pull effect, rather like a musical accordion. The P-waves cause particle oscillation parallel to the direction of propagation causing temporary lengthening and shortening of the material, which is why these waves are also known as longitudinal waves.

P-wave propagation is controlled by competence of the rock through which it is travelling. A rock that is weak, highly fractured or very compressible (i.e., soft) will impact the ability for compressional waves to effectively pass through without large amounts of energy being attenuating quickly. P-waves can travel through both matrix and fluid-filled pores of rock in the subsurface. The equation for the velocity of P-waves can be written using two different equations with one using elastic moduli and one using Lamé's parameter, but both have a density term:

$$V_P = \sqrt{\frac{\left(K + \frac{4}{3}\mu\right)}{\rho}}, \qquad (2.12)$$

$$V_P = \sqrt{\frac{\lambda + 2\mu}{\rho}}, \qquad (2.13)$$

where K and μ are the bulk and shear moduli, respectively, λ is the first Lamé's parameter and ρ the density of the material.

The other type of body wave we are interested in are secondary (S-wave) or shear waves which, as their name suggests, travel in a wave-like motion causing shear rotation (i.e., twisting) movement on the particles they pass through (Fig. 2.4). Because the particle motion is

Fig. 2.4 Schematic representation of the propagation of body waves in the subsurface

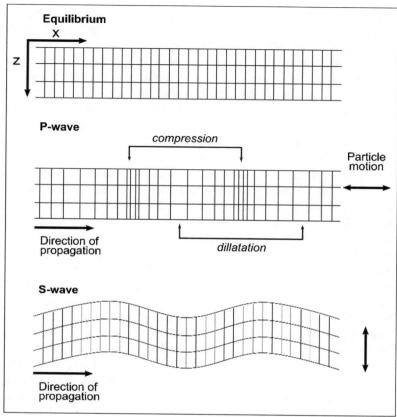

described as transverse, these waves are also known as transverse waves. S-waves are slower than P-waves, at approximately half the speed of the related P-waves. As mentioned, fluids are not able to retain shear stress and therefore S-waves cannot effectively travel through them. This has been extremely important in earthquake seismology as the inability of S-waves to travel through liquids or exceptionally low viscosity material led to the discovery and current definitions of the layer boundaries that make up the interior of our planet.

Within the energy industry, seismic surveys tend not to produce source-based S-waves because for offshore acquisitions they cannot propagate through the water column and on land they are difficult to consistently generate, as well as requiring specialized equipment to record. The equation for the velocity of S-waves is as follows:

$$V_S = \sqrt{\frac{\mu}{\rho}}, \qquad (2.14)$$

where μ is the shear modulus and ρ the density of the material.

Surface waves are typically responsible for the large ground movement in earthquakes and the ensuing destruction to infrastructures. The propagation of surface waves is directly guided by boundaries in the subsurface geology as the waves travel directly along interfaces between different velocity layers. The thickness of the velocity layer above and below an interface, controlled by geology, controls the maximum wavelength and velocity these surface waves can attain. The waves show little decay in energy as a function of attenuation when propagating along an interface. Consequently this type of wave exhibits extremely high amplitudes close to the ground surface which makes them a particular issue for onshore seismic surveys. Surface waves are the main cause of a phenomenon known as "ground roll" observed in recorded seismic data. They are removed during seismic processing as they give little information about the subsurface properties. If surface waves are not travelling along an interface their amplitude decays exponentially and therefore cause few issues in recorded seismic data.

Surface waves can be divided in two main types. Firstly, Rayleigh waves[5] produce a particle motion that at the surface can be described as elliptical and retrograde. This can be thought of as an ellipse in a vertical plane whereby most of the particle motion is vertical and the waves do not exhibit any perpendicular or transverse motion. Raleigh waves are characterized as being very low-frequency waves and the particle movement is affected by depth increase as well as the wave energy. The majority of ground roll recorded in an onshore seismic survey is made by these waves.

The second type of surface wave are Love waves[6] which exhibit a horizonal transverse particle motion that is parallel to the free surface. They are the result of the interference of many different S-waves guided by a low-velocity elastic layer on top of a half-space layer. They are low-velocity and, like Rayleigh waves, an increase in depth effects their amplitude which leads to decay.

2.1.4 Seismic Waves and Rays

So far, we have focused on describing the effect of propagating seismic waves on the particles within the subsurface, which is the criteria to classify the different types of seismic waves. In this section, it is important to consider the large-scale understanding of how waves propagate through the subsurface.

Seismic surveys are designed to image the subsurface by generating a pulse of seismic energy from a source location and determining, at the same or another location in the vicinity, the time interval between the initial energy pulse and the arrival of the seismic waves, which have been reflected and refracted from the velocity discontinuities (i.e., geological boundaries) in the material they are transmitted through. The focus of this section is exclusively on P-waves and S-waves. Below, we introduce the key principles and laws that underpin modern seismic wave and ray theory, many of which originate from the study of optics involving wavefronts and ray paths.

The Huygens principle[7] is a good starting point when discussing the propagation of waves. Huygens hypothesized that when considering a primary wavefront each point on that wavefront acts as secondary source point and generates energy to form a new wavefront. The new wavefront is therefore spatially defined by this secondary front (Fig. 2.5). A practical experiment to prove this hypothesis uses the behaviour of light hitting a small slit which then fans outward on the other side, known as diffraction.

We can visualize this concept by thinking about a pebble being dropped into a very still pond and the ripples (i.e., waves) that propagate out from the point where the pebble enters the water. The concept is that every point on a wavefront expands and moves away from the source point and acts like as new source of wave energy with matching phase and frequency.

Huygens' work helps us understand how secondary sources form and allow the spreading of reflected wavefronts. To further develop our understanding of wavefronts and how they interact with each other we need to consider the Fresnel zone.[8] The Fresnel zone is concerned with explaining the horizontal boundary of seismic waves upon a velocity interface, and results

---

[5] Rayleigh waves were introduced in an 1887 publication by Lord Rayleigh (1842–1919), a British scientist who worked in the field of physics and was a Nobel laureate.

[6] named after the English mathematician Augustus Edward Hough Love (1864–1940), after he won the Adams prize in 1911 for work on elastic waves.

[7] Named after the Dutch physicist Christiaan Huygens (1629–1695) whose achievements in mathematics and physics make him one of the greatest scientists of all time. In his book, *Traité de la Lumière* published in 1690.

[8] Named after the French physicist Augustin-Jean Fresnel (1788–1827).

Fig. 2.5 Schematic representation of the Huygens principle

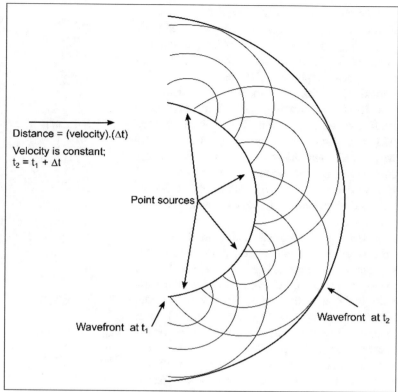

Distance = (velocity).(Δt)

Velocity is constant;
$t_2 = t_1 + \Delta t$

Point sources

Wavefront at t_1

Wavefront at t_2

from the research on optics and wave theory of light built on that of Huygens and Young. This leads to an explanation of diffraction of light waves which is transferable to seismic waves.

The Fresnel zone describes where waves travelling from a source point encounter a reflector in the subsurface, which leads to interference. The waves hit the reflector and can be reflected back in an infinite number of reflection ray paths that all interact with each other. Reflected waves will interfere with each other constructively where travel paths vary by less than half a wavelength of the initial reflected arrival. This constructive interference can be thought of as a summation that increases the reflected signal. The area where the constructive interference occurs is referred to as the first Fresnel zone and defines the horizontal resolution of the seismic data (Sect. 2.2.6).

2.1.5   Geometry of Reflected Rays

When a seismic source releases energy into the subsurface the energy travels in a variety of directions following complex paths. Conceptualizing seismic energy travel in terms of single vectors or rays means that they have a direction of travel and magnitude, which helps simplify the understanding of the seismic energy's path between source and receiver. We move now to examine the reflection and refraction of seismic ray paths.

The subsurface can be thought of as a series of layers that vary in physical properties, where areas of change affect how seismic energy travels through the layers. An area where one layer ends and another begins is called an interface and it represents a change in the rock properties. The two physical properties we are most concerned with are velocity and density, which vary relative to each other. Having some prior knowledge of the geology in an area and integrating it with the observed interface responses from the seismic data is very important. This integration is also critical for workflows in seismic processing, velocity model-building and seismic inversion. At these interfaces, some of the seismic energy will be reflected to the surface, whereas some will be refracted and transmitted deeper into the subsurface. The magnitude of contrast above and below the interface will affect the ratio of reflected and refracted energy of the seismic wave.

The ray path represents a simplified model of how wave energy propagates. An example of a ray path can be seen in Fig. 2.6 which includes a series of horizontal reflectors, positioned at variable depths with the source directly above. The reflector represents a change in V_P and/or V_S and density (ρ) large enough to result in some of the source energy being reflected back up to the surface-based receivers. However, not all the energy will be reflected as some will be refracted and transmitted further down towards the deeper layers. The time it takes for a pulse of seismic energy or source shot energy to travel down to each reflector is known as the one-way travel time (OWT) and the journey down to the reflector and back to the surface is called two-way travel time (TWT). If the reflected wave travels directly back to the source position this is described as a normal incidence reflection (Fig. 2.6).

The contrast in velocity and density of layers above and below an interface are important for the ratio of reflected/transmitted and refracted energy. In the normal incidence case, the amplitude of the reflected energy is determined by the P-impedance, which is defined as the product of the P-wave propagation velocity and density across the layer interface:

$$I_P = V_P.\rho. \qquad (2.15)$$

The fraction of reflected amplitude caused by the P-impedance contrast at the interface is measured by the reflection coefficient (R_{pp}):

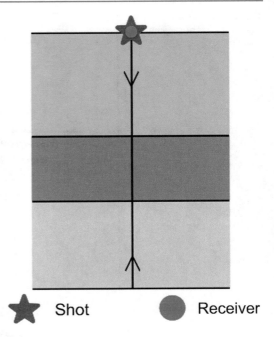

Fig. 2.6 Schematic representation of seismic energy travelling in the vertical direction and being reflected at normal incidence

$$R_{pp} = \frac{I_{P_2} - I_{P_1}}{I_{P_2} + I_{P_1}}, \qquad (2.16)$$

where the subscripts 1 and 2 refer to the mean above and below the reflection interface. A stronger reflector will have a large reflection coefficient and represent a large contrast in acoustic impedance at the interface, resulting in a high amplitude event on a seismic section due to the large amount of reflected energy. For a reflection to occur, the contrast between the elastic properties must be sufficient at the interface by either moving from a low P-impedance into a high P-impedance layer, or vice-versa. As such, we have both positive and negative reflection coefficients to signify whether we are going into something of higher P-impedance or lower P-impedance at the interface. The sign of the reflection coefficient is directly related to the phase of the wavelet at the event and how the amplitude of the reflection is recorded as either a negative amplitude or a positive amplitude.

P-wave attenuation is an additional property that is relevant when considering P-impedance contrast and reflected and refracted/transmitted energy. As the P-wave energy spreads away from the source and is transmitted through the subsurface, it only has a finite amount of energy. The depletion of this energy will lead to a decrease in the amplitude of the reflection events and the ability to produce strong refractions/transmissions. As the wavefront energy spreads out from the source origin in three dimensions, like a sphere, it enlarges and travels further from the source. The energy begins to decrease per unit area as it covers more area and is spread thinner. The increase in area is the square of the distance from the source and so the amplitude decreases inversely with the distance. This is often referred to as spherical spreading, or spherical divergence, and is dealt with in seismic processing by applying a gain function to compensate for its effect. Something else to consider is that where interfaces lead to very large reflection coefficients the amount of energy available for refraction/transmission will be reduced. This is common at geological features such as unconformities, salt and where clastic rocks meet either carbonate rocks or igneous basement. These geological discontinuities tend to produce very clear and strong reflection events in seismic data, providing a good geological boundary for interpretation of the geology in the area. However, this also means that the seismic imaging potential beneath these events is greatly reduced because not enough seismic energy remains to penetrate further below. In the North Sea, the classic example of such a geological discontinuity is the base cretaceous unconformity. These geological discontinuities are commonly rough and uneven boundaries which contribute to the reduction in transmission of energy beneath them (i.e., wave scattering).

Further complexity needs to be added to the normal incidence model to truly explain ray paths as they travel through the subsurface. Huygens' work shows that the ray fronts recorded for a single reflection event are not from a single point in the subsurface. The angle of incident is not always perfectly vertical when a seismic ray hits upon a reflector event, which means the corresponding reflection does not return straight back to the source point. Additionally, the magnitude of difference in velocity and density above and below the interface will influence the reflection and refraction angle.

Snell's law[9] describes the relationship between the travel path and angle from normal incidence that a ray takes while crossing a boundary or interface of separation and, furthermore, describes the paths of reflection and refraction coming from the contact with the interface. The incident and reflected angle are calculated to be identical, and the refracted (transmitted) angle is a function of incident angle and P-impedance contrast magnitude (Fig. 2.7):

$$\frac{\sin\theta_0}{\sin\theta_3} = \frac{V_1}{V_2},   (2.17)$$

where V are the media velocities and θ are the incidence and reflection angles (Fig. 2.7). It is possible to predict the angle of reflection and the angle of refraction given a known incidence angle of the down-going ray path. Another phenomenon that occurs when non-vertical incidence angles are considered is the generation of S-waves at the interface (Fig. 2.7) by a process of P-wave conversion, which occurs when a proportion of the reflected or refracted P-wave energy is mode converted into S-waves. We can utilize this phenomenon to generate S-waves from a P-wave source in a seismic survey, provided that the receivers are set up to measure these S-waves we gain additional useful data for describing the subsurface.

We now have the situation of down-going ray paths with variable incident angles that lead to reflections and refraction, which are not normal incident and will not return to surface at the source position. To record the reflected energy, the receivers (i.e., geophones or hydrophone; Chap. 3) are spaced at offset distances away from the source. We are now able to record seismic reflections that have travelled through the

---

[9] Snell's law was proposed in 1621 by the Dutch astronomer and mathematician Willebord Snell and it was used in the work carried out by Christian Huygens.

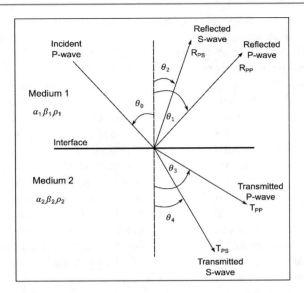

Fig. 2.7 Schematic representation of energy partition at an interface between different geological media

subsurface in a large range of different ray paths, as well as rays that have been refracted to much greater depths in the subsurface. Consequently, having a much more complex set of ray paths to deal with during seismic data processing, but that provide far more information about the Earth's layers.

The illustration in Fig. 2.8 is an overly simplistic representation of the possible multiple ray paths and reflection angles. When we introduce multiple layers, the ray paths travelling through them interact with multiple reflectors with non-normal incident angles, which cause a huge set of reflections and ray paths. Additionally, the geology may not be continuous and may vary in thickness and velocity (Fig. 2.9).

The complexity of the subsurface gives rise to unwanted ray paths called seismic multiple reflections or, simply, multiples. Multiples often represent the largest challenge for processing a seismic data set (Sect. 4.1) and in interpreting the image that comes from it. The ray paths from multiples are still recorded along with the other ray paths but it is the journey they have made through the subsurface that delivers false reflection amplitudes, as well as false information

about the location of the interfaces. Multiples can also mask and interfere with the image of primary reflections making subtle interpretation impossible. Some multiples are easy to identify given either the very fast or slow travel time compared with the primary arrivals at the receivers.

2.2 Fundamentals of the Seismic Survey

In this section, we briefly introduce some fundamental concepts related to seismic-based geophysical surveys. These concepts are important for the geoscientist to understand as they underpin many of the decisions made during interpretation and data analysis. The topics in this chapter are linked to further discussions in the following chapters.

2.2.1 Geophysical Survey Signal

The general purpose of a geophysical survey is to measure the strength of various physical properties of the subsurface, whether it be the strength

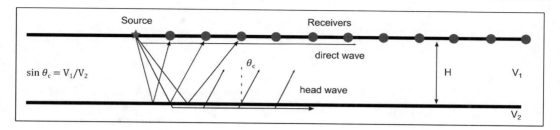

Fig. 2.8 Schematic representation of reflected and refracted rays originated at the source and detected at the receiver positions. θ_c is the critical angle

Fig. 2.9 Schematic representation of a complex geological environment where the ray paths travel from the source downwards and are reflected at the interfaces between different geological layers upwards, and are detected by the receivers

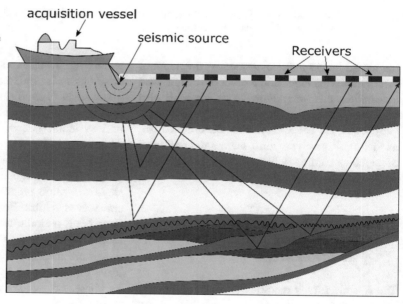

of a magnetic field, gravity field or acoustic waves, to either understand its spatial variation or variation in strength over a set period. The results from such a survey can be plotted on a graph with time, or distance, on the x-axis and magnitude on the y-axis. The plot will be a representative shape referred to as a waveform of the data collected, which can be easily visually interpreted. Figure 2.10 shows an example of two fictional survey results one for gravity and one for ground velocity from a seismogram.

The recorded survey shown in Fig. 2.10 produces a smooth continuous function that contains information about both the large and small variations in the signal for distance with gravity and a period of time for the seismogram. The goal of any survey is to record as much data as possible

and, as a general rule, the more data collected the greater the ability to understand the details of variability, which lead to a more comprehensive interpretation. Modern seismic surveys are capable of recording huge amounts of data for this very purpose. A marine seismic survey is technically able to record data continuously by ensuring the record length (i.e., time the receiver is listening) is longer than the time between shots. Consequently, huge amounts of data are recovered compared to a non-continuous acquisition, where after a fixed period the recording system is cycled before the new shot is triggered, thus omitting to record some returning signal. To reduce the amount of data recorded, the shot interval can be shortened to make continuous recording easier, but this would be detrimental to

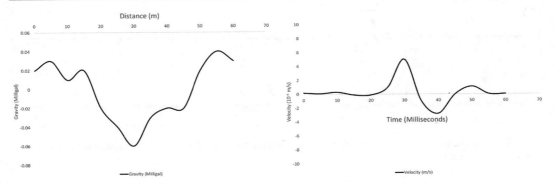

Fig. 2.10 An illustrative example of a gravity profile showing field variability with distance (left) and a seismogram showing the particle velocity variation with time

the record length and, consequently, the depth of information obtained. Some modern seismic land acquisition that uses wireless technology also continuously records the signal picked up by the geophones. However, since the data are recorded as digital all receivers must have a defined level of signal sampling which defines the ultimate rate at which data are recorded.

2.2.2 What Does the Event on a Seismogram Represent?

The seismic survey records the oscillation of the ground particles caused as the seismic energy transfers through them. Remember that the wave motion is a transfer of energy from one point to another, with often no lasting distortion of the media particles. The motion of a particle's velocity can be thought of as a vibration, for example when a tuning fork moves back and forth from its position of rest. The true behaviour of the particle's oscillation is highly complex because the wave interacts in three dimensions and the structure of the subsurface is heterogeneous; but to understand how the seismic data relates to the geology it passes through only requires an understanding of the reaction of the particle's range of movement and how these change with time.

One way to represent the repetitive movement of an object is to plot it graphically as a sine

wave (Fig. 2.11). A recorded seismic wave on a seismogram is continuous in nature and exhibits a sinusoidal waveform. Sine waves are a common representation of the changing state of natural systems (i.e., wind and waves and light waves) and this also applies to the particle motion in media caused by seismic waves propagating through it. Cosine waveforms also share the same repetitive cyclic properties that sine waves exhibit. Both waveforms are useful at expressing periodic movement shown as amplitude variations along a time axis. Sine and cosine waves are identical except that the cosine wave leads the sine wave by 90° (i.e., $2/\pi$ radians) in its phase. So, whereas a sine wave starts at an amplitude of zero, a cosine wave starts at an

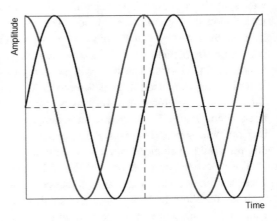

Fig. 2.11 An example of a sine (black curve) and cosine (blue curve) waves

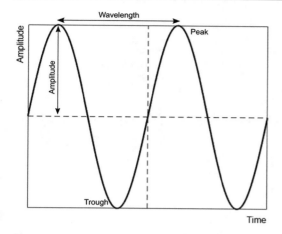

Fig. 2.12 Basic components of a repeating waveform

amplitude of one. Figure 2.11 compares a sine and cosine wave.

An interesting defining characteristic of the shape of a sine or cosine wave is that the greater the amplitude the slower the rate of change. Conversely, the rate of change is fastest when approaching zero magnitude. The uniform and repetitive nature of these waves allows their properties to be well understood as they do not exhibit random variability with time. The basic measurements of a repeating wave form, such as amplitude and wavelength are illustrated in Fig. 2.12.

2.2.3 Relationship Between Particle Movement and Sine Wave

To visually relate what the sine or cosine waves from a geophysical seismic survey show about the physical movement of the media particle (i.e., caused by seismic wave interaction), we use a simplified model of a circle (Fig. 2.13). Measurements are made of its diameter at different locations on its perimeter from a central point to represent particle movement. The diameter of the circle represents the maximum physical displacement of the particle from its resting position (i.e., centre point of the circle) caused by the interaction with the seismic wave (Fig. 2.13). We can use these lines to superimpose triangles onto

the circle. A sine function plots the height of the triangles from the horizontal x-axis and divides by the radius of the circle, which is represented by the hypotenuse of each triangle (dashed line in Fig. 2.13). The sine wave plotted on the graph with the x-axis in degrees shows how each triangle's height is plotted to create a point that makes up part of the sine wave. Figure 2.13 shows that the sine or cosine wave can be explained by three main components: the amplitude limits of the peaks and trough, the phase shift that the data have and the frequency content.

The amplitude of the waveform is related to the magnitude of the movement of the particle which is represented by the diameter of the circle model (Fig. 2.13). If the line that rotates around the circle were to decrease in length the amplitude of the waveform would also decrease, which would result in the peaks and troughs of the sine wave becoming smaller.

The faster the line rotates around the circle the higher the frequency of the resulting sine wave. Frequency is measured in terms of complete cycles within a specified time period (e.g., 1 s). Hertz[10] is the international system of units for frequency, with 1 Hertz (Hz) being equal to one cycle per second. One important distinction between frequency and cycle period is that the cycle period is the time taken for a particle to make one full vibration cycle, but the frequency is how often this event happens and is the measure of rate. This leads to a relationship where the cycle period is the reciprocal of the frequency with the reverse being true:

$$T = \frac{1}{f},\qquad(2.18)$$

$$f = \frac{1}{T},\qquad(2.19)$$

where T is defined in time and represents the period of the wave and f is the frequency in Hz.

---

[10] Named after Heinrich Rudolf Hertz (1857-1894), a German physicist, who proved the first definitive evidence for the existence of electromagnetic waves.

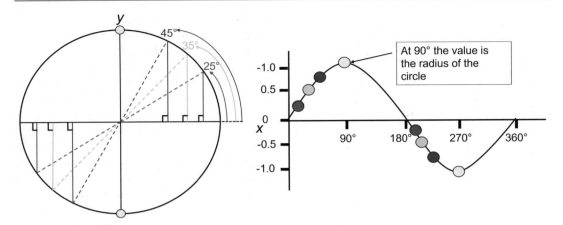

Fig. 2.13 How a line from the centre of a circle to its perimeter moving in a circular movement can represent the amplitude variations of a sine wave

So far the sine waves we have looked at have a single frequency, but in reality there is a large range of frequencies associated with a single seismic wave and these are called harmonics. Being able to measure the spatial repetition of the wave will give the wave cycle, or more commonly known as the wavelength. The distance from one positive peak to another, often referred to as crest to crest, is one full particle oscillation. Using the example in Fig. 2.13, one full particle oscillation would represent a full revolution around circle where we have plotted around the full 360° ($2\pi\, radians$). If we were to replace degrees on the x-axis with time, this would give the ability to calculate the period of the oscillation using the following formula as an expression of time, which would allow us to calculate how many wave cycles per second to expect:

$$T = 360\pi/\omega, \qquad (2.20)$$

where ω is the angular frequency.

Finally, the waveform phase of the seismogram is often not well understood. Phase is generally discussed when comparing two waveforms. The frequency, period cycle and even amplitude can be the same between the waveforms, but the phase of a waveform can vary from another waveform. Figure 2.14 shows an example of two waveforms that are out of phase by 90°. If the peaks and troughs of the two

waveforms are not aligned at the same time they are considered to be out of phase. Phase can be measured in several ways such as distance, time or degrees. Using time on the horizontal axis can show what level of particle movement is at time zero and also how quickly the maximum amplitude is reached.

Seismic data are commonly processed to be zero-phase. A zero-phase wavelet is symmetrical around time zero and shows energy before time zero, which is not possible in reality. A more representative phase is observed in a minimum phase wavelet. Zero-phase processing is driven mainly by the needs of the interpreter where a zero-phase data set gives the ability to pick changes in reflectivity in the middle of peaks and troughs, making life a lot simpler. The process of zero-phasing is used because the seismic source wavelet in the survey will have a minimum phase character. At time zero, the source has energy and therefore no amplitude or phase, but this changes as soon as the shot is fired and greatly increases and then decreases in a very short period of time. For the interpreter, the minimum phase wavelet will, when compared to a perfectly zero-phase wavelet, have degrees of rotation, meaning it is not perfectly symmetrical around time zero.

The model shown in Fig. 2.15 focuses on explaining small-scale particle oscillations. To consider bigger volumes of media, with more

Fig. 2.14 Two waveforms that are out of phase by 90°

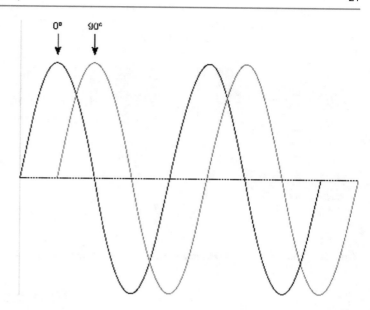

particles, allows the effects of physical distortions to be upscaled while remaining relatable to the peaks and troughs in a sine wave. Figure 2.15 shows the repeating pattern of compression and rarefaction as a P-wave distorts particles in the media it is travelling through. The area of particles undergoing maximum compression are represented by positive peak amplitude values in the sine wave. Areas undergoing maximum rarefaction are represented by the negative trough amplitude values.

Areas of media undergoing compression during wave propagation experience high pressure because the distance between media particles is reduced and they are forced closer together. At the point of maximum pressure the particles are static, but the particles around them are getting closer and applying pressure. The opposite is found in areas of rarefaction which have low pressure because the distance between media particles is made greater as they become separated. At the point of minimum pressure, the particles are moving the most. The equation used to describe the displacement of sound waves ($s(x, t)$) resulting from the physical effect of the sound wave going through the media is defined as

$$s(x, t) = S_m \cos(kx - \omega t + \varphi), \qquad (2.21)$$

where kx is the angular wave vector, ωt the angular frequency, φ the phase shift and S_m the amplitude of particle displacement. The change in media particle pressure (Δp) due to the sound wave can be described as

$$\Delta p = Bks_m \sin(kx - \omega t + \varphi), \qquad (2.22)$$

where Bks_m is the amplitude of particle pressure.

If you were to compare a cosine and a sine wave representing particles in a media undergoing compression and rarefaction using Eq. 2.2, an area undergoing maximum pressure would show a corresponding zero-amplitude response on the cosine wave for particle displacement and vice versa for an area undergoing minimum pressure.

So, when one or more of these variables alter the appearance of the waveform what does this tell us about the changes in particle motion being recorded in the seismogram? Fig. 2.16 aims to explain this effect. The dashed lines move around the circle in an anti-clockwise direction and as the line continues to move towards a vertical position within the circle, the amplitude values in the sine wave increase until the maximum

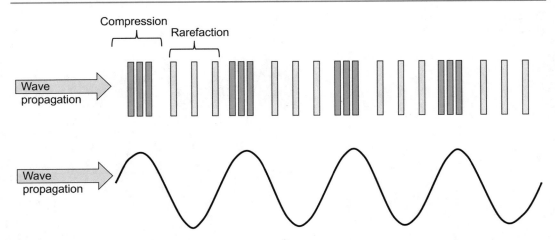

Fig. 2.15 The relationship between a P-wave's peaks and troughs with effect the wave has on particles in the media

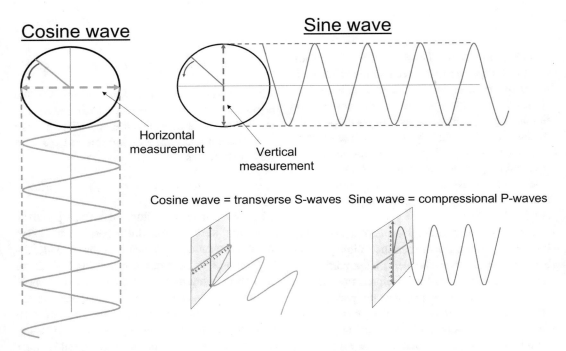

Fig. 2.16 Schematic representation of the relationship between a cosine and a sine function

amplitude value is reached when the line is at 90° from the x-axis. As the line continues to rotate past 90° from horizontal and eventually goes below the x-axis line, the amplitude values of the sine wave become negative with the maximum negative amplitude being at −90° to the x-axis line. The circle analogy also works for plotting a cosine wave, however, rather than measuring the height from the vertical y-axis it is measured from the horizontal x-axis. The cosine wave is plotted 90° rotated compared to the sine wave (Fig 2.14) and, for this reason, the sine wave can be used to represent compressional wave motion (P-waves) and the cosine wave to represent transverse wave motion (S-waves).

2.2.4 The Fourier Transform

Geophysical signal processing and seismic analysis often require an estimation and isolation of the source pulse produced by the seismic source (Chap. 5) to understand the relationship between the reflectivity events that make up a recorded seismic trace and the geology. The source pulse, or wavelet, is a short, high energy, discrete waveform with a range of frequencies that have a defined start and end time embedded in the continuous waveform, which is recorded during the geophysical survey. Continuous waveforms have no defined start and end time, so it is not possible to extract a wavelet from seismic data in this form and requires a process called the Fourier transform.

Any waveform that repeats itself at a fixed period with the exact same shape given by its amplitude and phase characteristic is referred to as a periodic waveform. An individual sine wave at a particular frequency from a geophysical survey certainly fits these characteristics. We know that a seismic signal is made of not just one frequency but many and can be represented as sine waves with different amplitudes and even phases.

A recorded waveform contains a range of frequencies referred to as the fundamental frequency and harmonics. Understanding and being able to visualize the frequency content of our seismic data becomes very important for seismic analysis because it points to the scale of features the data can resolve and why they might appear as they do in a seismic image. In the time domain, we see the waveform as a function of particle displacement amplitude varying with time; but in the frequency domain it is expressed in terms of amplitude and phase for a particular frequency.

The time domain display allows us to interpret the shape of a signal and how it varies with time. In the time domain, it is difficult to know how many frequencies are in the data as each individual sine wave has a small amplitude. In the frequency domain, a single waveform can be plotted as a point for both amplitude and phase (Fig. 2.17a). Multiple waveforms for the entire signal are often plotted together for amplitude. When these waveforms are combined, we are able to measure the density variation of amplitude and identify areas of low and high frequencies (Fig. 2.17b). For these plots, the vertical axis is often shown as power in decibel (dB) to measure the acoustic frequency strength. Viewing the waveform in the frequency domain allows us to see not only how many frequencies make up the signal (i.e., frequency range) but how strong the individual frequencies are relative to others within the data range. This variation in frequency content and distribution comes about because when waveforms interact with a physical object (i.e., a geological boundary) its frequency spectrum becomes altered. As a rule of thumb, high frequencies are lost first as the wave travels through the subsurface because they have shorter wavelengths which attenuate quicker than the lower frequencies (i.e., the Earth acts as a low-pass filter). Understanding this variability is key

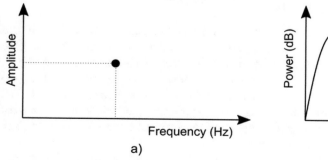

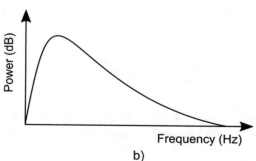

Fig. 2.17 **a** single waveform plot in the frequency domain and **b** all the waveforms of a seismic signal

because it may impact what value we are able to get out of the seismic data at a particular depth in the subsurface.

The Fourier transform is a mathematical transform, that allows the data to be converted from the time domain into the frequency domain. The signal in time can then be split out into different frequencies from that waveform and combined together into a complex periodic waveform that expresses the variabilities of the individual sine wave's frequencies.

The Fourier transform uses a single sine wave with a single frequency, generally on the lower end of the range, referred to as the fundamental frequency. The fundamental frequency is added to with a linear combination of sine and cosine waves which are integer multiples of the fundamental frequency, referred to as harmonic frequencies. The harmonics are kept to the upper frequency limit which is suitable for the recorded data.

If the signal $g(t)$ is periodic, infinite and with period T then the signal might be represented by a Fourier series:

$$g(t) = \frac{a_0}{2} + \sum_{n=1}^{\infty} \left[a_n \cos\left(\frac{n\pi t}{L}\right) + \sin\left(\frac{n\pi t}{L}\right) \right],$$
$$(2.23)$$

where

$$a_0 = \frac{1}{L} \int_{c}^{c+2T} g(t) dt, \qquad (2.24)$$

$$a_n = \frac{1}{L} \int_{c}^{c+2T} g(t) \cos\left(\frac{n\pi t}{L}\right) dt, \qquad (2.25)$$

$$b_n = \frac{1}{L} \int_{c}^{c+2T} g(t) \sin\left(\frac{n\pi t}{L}\right) dt. \qquad (2.26)$$

In complex seismic traces, as the recording length (t) increases it takes longer for the signal to repeat and, in the case when T becomes infinite $g(t)$ no longer repeats so we use:

$$G(f) = \int_{-\infty}^{\infty} g(t) e^{-j2\pi ft} dt, \qquad (2.27)$$

$$g(t) = \int_{-\infty}^{\infty} G(f) e^{j2\pi ft} df, \qquad (2.28)$$

where $G(f)$ is the Fourier transform of $g(t)$ and $g(t)$ is the inverse Fourier transform of $G(f)$. $G(f)$ and $g(t)$ are a transform pair:

$$G(f) \leftrightarrow g(t). \qquad (2.29)$$

When combined the complex periodic waveform appears as a product of the fundamental and harmonic variations in frequency. Amplitude and phase are also important and are expressed clearly in the shape of the complex periodic waveform. Harmonics that are out of phase from the fundamental frequency are reflected in the shape of the periodic waveform. Figure 2.18 shows an example of the summing of sine waves at different frequencies to form a complex periodic waveform. The series of components combined to make the periodic waveform are referred to as a Fourier series. The periodic waveform has a square appearance but is an approximation of the input sine waves. If we were to isolate a single full sequence from the periodic waveform this would be referred to as a transient waveform and represents a short burst of energy over a specified time period, which allows us to get closer to the generated seismic wavelet.

2.2.5   Sampling

A seismic signal like that shown in Fig. 2.10 can appear as a continuous function in the time domain and is referred to as an analog signal. When a geophysical survey records data, the continuous waveform is digitally sampled. The regularity or frequency of this sampling is very important to correctly represent the waveform shape.

The principle of digital sampling of a continuous waveform is reasonably straightforward.

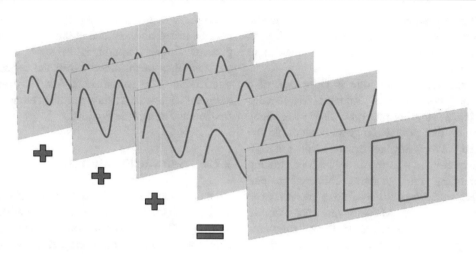

Fig. 2.18 Summing of multiple sine waves at different frequencies to produce complex periodic waveform

The digital seismic signal is composed by many regularly sampled fragments of the continuous waveform detected at each receiver. Each sample is a single measurement of a signal magnitude at a particular time. The idea is that the smaller the sample interval the more likely the true variations and shape of the waveform can be identified. However, as a consequence more signal data samples need to be stored and dealt with during the data processing stage. The digitized form of the continuous waveform should capture both the large and small variations. Even today, with easy access and availability to large digital storage services, the costs and associated backups can result in substantial expenditure.

The sampling frequency relates to the number of points per unit of measurement (i.e., time or distance). For a conventional marine seismic survey in the energy industry (not including ultra-high-resolution seismic surveys used, for example, in geotechnical surveys) 2 or 4 ms is a common sample rate, with onshore surveys being 1 ms or less for very shallow surveys.

Once the signal has been digitized, it is reconstructed back to the continuous signal. Sampling a continuous function at discrete values risks the loss of important information, however it can be negated by ensuring that the sampling frequency is higher than the highest frequency component of the function being sampled. For a

sine wave, a minimum of two discrete sample values per period should make reconstruction achievable. This is explained by one of the most important laws in signal processing: the Nyquist Theorem. The theorem defines that the highest frequency able to be represented accurately is exactly one half of the sampling rate. The Nyquist frequency[11] gives us the limit of frequency we can sample:

$$f_{NQ} = \frac{f_0}{2}, \qquad (2.30)$$

where f_{NQ} is the Nyquist frequency and f_0 is the sampling frequency.

A simple worked example would be a waveform as a continuous sine wave being sampled every 4 ms (sample interval) which would equate to 250 samples per second (1000 ms = 1 s) or 250 Hz. Based on the sampling rate we are only able to correctly sample up to 125 Hz, which is half the Nyquist frequency (f_{NQ}). This will have an effect on the dynamic range of the data set.

The dynamic range of a digitized data set (expressed in decibels) refers to the ratio between the largest and the smallest measurable amplitude. Data with a high dynamic range is an indication that sampling has picked up lots of

---

[11] Named after Harry Nyquist (1889–1976) a Swedish Physicist and Engineer.

amplitude variation from the continuous wave-form and gives the next step of wave digitization a better chance to represent the maximum variability.

The recorded seismogram is composed of a range of information at different frequencies. Choosing a sampling rate for a single-frequency component that provides a good outcome in the digitized wave is straightforward. Yet, for multiple frequency ranges from low-high calculating the sampling rate is more challenging, and incorrect sampling will lead to a phenomenon known as aliasing. Aliasing can occur during sampling and digitization of the waveform, when the sampling rate is too low, which leads to higher frequencies (above the Nyquist frequency) being incorrectly represented at their true frequency. These high frequencies become mixed back or "folded in" with the correctly sampled lower frequencies below the Nyquist, which leads to distortion effects and artefacts in imaging. There are two simple solutions commonly applied to minimize this issue. First, is to ensure that the sampling frequency is always a minimum of two times higher than the highest frequency in the sample function. This may be impractical for seismic surveys as the high

sampling will result in a substantial increase in data collected and associated costs. The second, perhaps simpler, option is to apply an anti-aliasing filter to the data before the process of digitization. The anti-aliasing filter is a simple low-pass filter with a sharp cut-off to exclude any frequencies above the Nyquist.

2.2.6 Seismic Resolution

Seismic resolution can be defined simply as the ability to distinguish between two separate features, or objects, that are close together. The higher the resolution the more detail we can see and interpret, which is important because of the inherent complexity, heterogeneity and small-scale variability of the subsurface. Although we may have a good understanding of geological processes and depositional environments from either outcrops, analogs or even well-logs and samples, we can never remove the unpredictability and irregularity of nature.

Resolution exists in two directions: horizontal and vertical. Both are important to interpretation and understanding the subsurface. Figure 2.19 shows an example of trying to interpret discrete

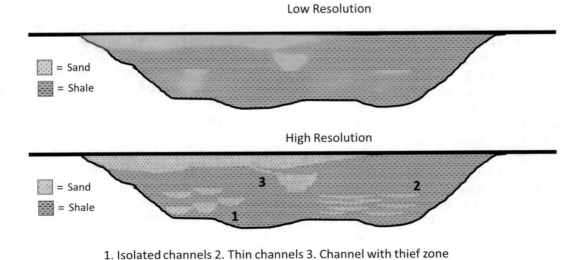

1. Isolated channels 2. Thin channels 3. Channel with thief zone

Fig. 2.19 The effect of resolution when interpreting small-scale geological features. The low-resolution does not allow to interpret the isolated features

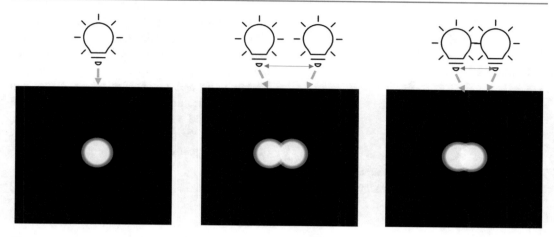

Fig. 2.20 (Left) Single light source produces a single aperture circle with diffraction. (Middle) Two light sources produce two aperture circles with clear destructive interference on the inner edges but still two clear circles. (Right) Two closely spaced light sources makes it difficult for the observer to resolve the representation of two light sources

deposition features when the resolution is poor. Features that look like large continuous channels can in fact be disconnected, which may impact their economic value as viable reservoir units. Disconnected sedimentary features are harder to develop and contain much less volume than anticipated. Furthermore, the seal may be compromised leading to lack of charge within the reservoir unit. These are important concepts when dealing with hydrocarbon or CO_2 reservoirs.

The first definition of resolution relates to the behaviour seen in optics. It was first defined by Rayleigh, and the principle is based on observing the intensity of light as passing through an aperture (Fig. 2.20). One might expect to see a whole circular aperture filled with an equal intensity of light. However, as the light passes through we observe an intense and well-formed central circle of light with a less defined, dimmer circle around the brighter core. This pattern is caused by diffractions where light from different areas of the aperture causes the constructive and destructive interference of light waves as they travel through it.

If we add an additional light source next to the first, we will then observe two small light circles overlapping in the centre as they come through the aperture. Basically, two representations of the

light sources with some level of diffraction and constructive and destructive interference between them cause the inner most edge to be lost. If we move the light sources closer together, we start to lose the ability to distinguish between the two circles and, therefore, lose the resolution to see both the separate light sources.

The observation from this optical resolution experiment shows that diffractions, constructive interference and destructive interference all play a role in effecting the visible resolution of travelling waves. The criterion established by Rayleigh is to define the peak-to-trough separation as the central-maximum to the adjacent-minimum time interval of a diffraction pattern as the limit of resolution. This means that two point-source objects are considered resolved when their separation is equal to or exceeds the peak-trough separation (i.e., half a wavelength) (Kallweit and Wood 1982). However, for this method of approximating resolution to be applied to seismic data it was revised in the form of the Ricker resolution criterion, which focuses more on the reflection coefficient event observed in seismic data. It refines the definition of the limit of resolvability to be when the waveform from two reflections has zero-curvature at its maximum amplitude. This gives a slightly different resolution limit of a quarter of the wavelength.

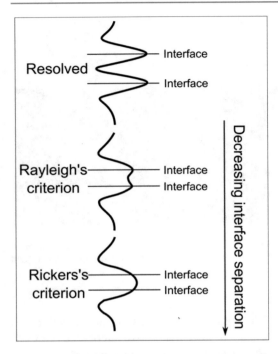

Fig. 2.21 Variation between Rayleigh's criterion and Ricker's criterion regarding the seismic resolution

Fig. 2.21 shows the comparison between the Rayleigh and the Ricker criteria.

A few other factors will be at play in the resolution of real seismic data. The level of noise within the data may obscure individual reflection events. Often the noise can lead to the event appearing broken up. The strength of the event is important because it will be difficult to distinguish if it has a weak amplitude. This may be a factor of the low impedance contrast, poor illumination or lack of suitable seismic energy present, especially below large contrasting layers or deep in the subsurface. In addition, complex highly dipping geology can have negative influence on the theoretical vertical resolution limitation.

Data sampling (Sect. 2.2.5) plays a part in what frequencies are present in the data. The larger the sampling frequency, the higher the frequencies present in the seismic data. This result is obtained directly from the Nyquist theorem (Eq. 2.29).

To consider vertical resolution in seismic reflection data, we think about defining the top and base of geological layers of our potential target. Also of concern is the overburden geological units and geometries where layers can become thinner and pinch out to form important features. Based on the Ricker limit of resolution, the bed we want to image must have a thickness of at least a quarter of the wavelength:

$$V_{res} = \frac{\lambda}{4}, \qquad (2.31)$$

where

$$\lambda = \frac{v}{f}, \qquad (2.32)$$

and λ is wavelength, v is the wave propagation velocity and f the frequency of the signal at the depth of interest.

Equation 2.31 shows that increasing velocity in the subsurface will negatively affect the resolution. Generally, we expect velocity and, consequently, the wavelengths to increase with depth. Attenuation of seismic frequencies is expected to occur as depth increases, with the shorter, higher frequencies being attenuated in the shallow subsurface leaving the longer lower frequencies to travel deeper (Fig. 2.22). As a result, the deeper features will need to be thicker for the top and base to be detectable with seismic reflection data. This aspect brings two important factors into play; the frequency content of the seismic and the wavelet size and character.

The frequency range of the seismic data ultimately determines the wavelength of the data, along with the velocity at the depth of interest. For the calculation of wavelength, a safe value to use is the "peak frequency" of the data at the depth of interest, which is defined using a frequency spectrum plot and choosing the frequency with the highest amplitude. This can be harder to do in modern data which has been deghosted and shaped because the broadband frequency spectrum has a flat top and a range of perhaps 15–20 Hz of frequencies at the highest amplitude (Fig. 2.23).

Fig. 2.22 Vertical seismic section illustrating the variation of frequency with depth

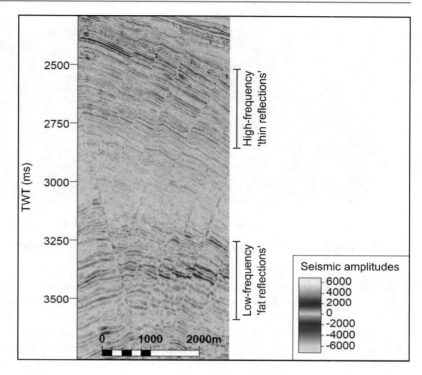

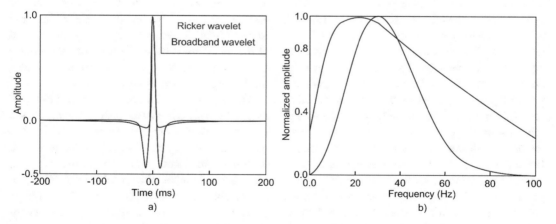

Fig. 2.23 Comparison between broadband and Ricker wavelets: a) in the time domain; and in the frequency domain

As previously mentioned, geological units can still be detected and even mapped if they fall below the quarter of a wavelength threshold, so this is not a hard limit applied to all data. The amplitude of a single peak or trough is actually a combination of much smaller variations in acoustic impedance. This is easily seen when comparing a well-log for acoustic impedance with the synthetic seismogram generated from the well convolved with a chosen wavelet. Geological features such as very acoustically soft gas sands or very acoustically hard cemented sand or carbonate stringer, commonly seen in the tertiary aged rocks from the central U.K. North

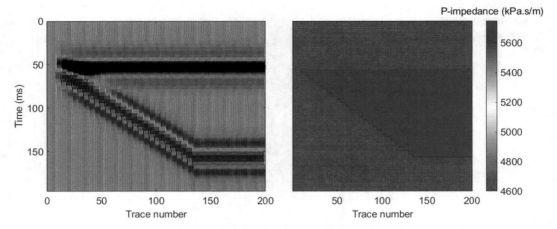

Fig. 2.24 Example of the tuning effect in a wedge model considering a Ricker wavelet with 50 Hz as central frequency

Sea, will contribute to the overall amplitude response of a reflection event, even though they are below the resolution of the seismic data.

Amplitude tuning is an effect related to resolution and amplitude variation caused by changes in thickness of two events. The standard analysis of tuning is done using a wedge model which synthetically generates the seismic response of a triangular-shaped wedge of two different facies (Fig. 2.24). The resolution shown by the model is controlled by a wavelet. The wedge is created by defining a simple model of acoustic impedance properties of the reservoir wedge and the non-reservoir in which it is encased. Where the wedge thickens, and is over a quarter of a wavelength, the model will show clear top and base reflections for the wedge, as well as internal reflections caused by sidelobes. As the wedge become progressively thinner the separation distance between the top and base event becomes smaller. Eventually, the one quarter of the wavelength distance is reached and we start to see interactions between events that represent the top and base of the wedge. This is termed the tuning thickness of the data and represents a thickness at which and below we are no longer able to get a clear reflection from the top and base of a seismic unit.

Before this point, we will observe interaction between sidelobes of top and base events which is usually destructive in nature due to their opposite polarities. This observation is often overlooked but is important to be aware of. The first effect that we tend to see is termed constructive interference, where the sidelobe of the base event starts to combine with the reflection of the top event to artificially increase its strength; referred to as amplitude brightening. As the wedges gets even thinner, the event of the base starts to interact with the opposite polarity event of the top. This is called destructive interference as, due to the opposite polarities, we start to see a weakening or dimming of the top event.

Tuning effects are a concern when looking at prospects with associated AVO (Chap. 6) and with geological units that become thin. In fact, sometimes, we rely on stratigraphic features such as up-dip pinch outs, where sands become thin, for hydrocarbon trapping mechanisms. The tuning is able to make both amplitudes brighten through constructive interference and deconstructive interference.

Seismic data are probably the best data set in terms of sheer horizonal coverage of large areas. However, when it comes to resolution in the horizonal direction we find that this is inferior to that of the vertical. This is mainly due to the waves from the source dispersing out and away from the source. The first Fresnel zone defines the theoretical value for the horizontal resolution of the data as its radius (R) can be defined as:

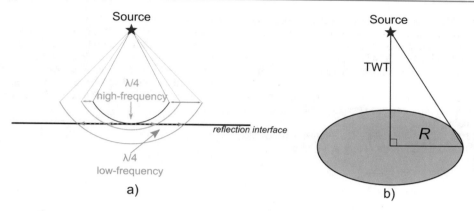

Fig. 2.25 **a** Illustration of the Fresnel zone for two different frequencies (low and high); **b** definition of Fresnel radius (*R*)

$$R = \frac{V}{2}\sqrt{\frac{TWT}{F}}, \qquad (2.33)$$

where V is the propagation velocity of the seismic wave, TWT is the two-way travel time and F is the frequency at a given depth (Fig. 2.25). In practice, the size of the Fresnel zone is affected by the depth of the reflector, frequency of the seismic at that depth and the propagation velocity of the rock.

Defining the Fresnel zone is a three-dimensional problem and historically posed pitfalls for two-dimensional seismic survey geometries. Reflection information from out of plane, which may represent quite different velocities, can falsely contribute to reflectors' shape and amplitude in the plane of the two-dimensional section. The process of two-dimensional migration results in the collapse of the Fresnel zone in one orientation along the survey line, while in other directions it still extends out beyond the plane of the two-dimensional survey line. This contamination may make in-plane reflections appear like false seismic anomalies or structures, and such interference has historically led to numerous failed wells. Beyond this zone, reflected waves tend to act both constructively and destructively and, effectively, cancel each other out so that the edges of the Fresnel zone defines the absolute horizontal limit of the reflection. In other words, a recorded reflection point is reflected from a wider spatial area defined by the Fresnel zone and not a single reflection point. Seismic data coverage of a target should extend a full Fresnel zone beyond its boundaries to allow seismic processing migration to pick a suitable aperture to recover the true amplitude value (Schleicher et al. 1997). For this reason, the geoscientist should be wary of interpreting features and amplitude anomalies on the very edge of survey areas and should check it is within the migration aperture.

References

Avseth P, Mukerji T, Mavko G (2005) Quantitative seismic interpretation: applying rock physics tools to reduce interpretation risk. Cambridge University Press, Cambridge

Chopra, S, Marfurt, KJ (2005) Seismic attributes—a historical perspective. Geophysics 70:3SO–28SO

Dobrin MB (1976) Introduction to geophysical prospecting, 3rd edn. McGraw-Hill

Goodway B, Chen T, Downton J (1997) Improved AVO fluid detection and lithology discrimination using Lamé petrophysical parameters; "λρ", "μρ", & "λ/μ fluid stack", from P and S inversions. SEG Technical Program Expanded Abstracts, 183–186

Kallweit RS, Wood LC (1982) The limits of resolution of zero-phase wavelets. Geophysics 47(7):1035–1046. https://doi.org/10.1190/1.1441367

Kearey P, Brooks M, Hill I (2002) An introduction to geophysical exploration, 3rd edn. Blackwell Science

Quakenbush M, Shang B, Tuttle C (2006) Poisson impedance. Lead Edge 25:128–138

Russell B, Hedlin K, Hilterman F, Lines L (2003) Fluid-property discrimination with AVO: a Biot-Gassmann perspective. Geophysics 68:29–39

Schleicher J, Hubral P, Höcht G, Liptow F (1997) Seismic constant-velocity remigration. Geophysics 62(2):589–597

Sheriff RE (1991) Encyclopedic dictionary of exploration geophysics, 3rd edn. Society of Exploration Geophysicists, Tulsa

Sheriff RE, Geldart LP (1995) Exploration seismology, 2nd edn. Cambridge University Press, Cambridge

Verm R, Hilterman F (1995) Lithology color-coded seismic sections: the calibration of AVO crossplotting to rock properties. Lead Edge 14(8):847–853

Zhou X, Ba J, Santos JE, Carcione JM, Fu L-Y, Pang M (2021) Fluid discrimination in ultra-deep reservoirs based on a double double-porosity theory. Front Earth Sci 9:649984. https://doi.org/10.3389/feart.2021.649984

Seismic Acquisition Essentials

Seismic data acquisition is performed to answer a specific scientific problem (e.g., the most favourable place to inject and store CO_2, the location of hydrocarbon reservoirs), whether this is to reduce uncertainty and increase knowledge for a particular area of interest in the subsurface or to more broadly increase general understanding of the internal structure of the Earth.

A seismic survey should be thought of as a scientific experiment and, indeed, practitioners in the field often refer to it as such. Seismic surveys can range hugely in complexity, duration and cost depending on the type of data collection method used and the size of area being surveyed. The simplest survey may comprise a group of university students with a hammer, a metal plate, a few geophones and a laptop to acquire a few short two-dimensional profiles. On the other end of the scale, a cutting-edge acquisition for the energy industry can cost tens of millions of dollars, involve multiple large acquisition vessels (i.e., offshore acquisition) or fleets of trucks (i.e., onshore acquisition) and up to a hundred people in various roles taking months to complete.

The parameters of the survey design and conditions in which it is acquired will play a large role in the suitability of achieving the project aims. Seismic data recorded in the field are termed raw data and are not suitable for the interpretation of the subsurface geology. How-ever, the QC and evaluation of these raw data is incredibly important because it makes sure that the recording equipment is working correctly and recording useful data for subsurface interpretation. If a man-made source is involved then timing, output and consistency of the source should also be QC'd. In large, modern land surveys the number of receivers can be so large that data might be acquired without field acquisition QC (i.e., 'blind shooting') due to the existing redundancy in terms of receivers. Even if a percentage of receivers is not properly functioning, avoiding field QC can represent a decrease in the costs associated with the acquisition, without compromising the quality of the data.

It is good practice for any geoscientist working with seismic data to understand the essential aspects of seismic acquisition, survey planning, acquisition QC, processing and interpreting seismic data. Having a basic technical understanding of how seismic data are acquired is important as it plays into why a final seismic data set looks the way it does. Project data are not necessarily a from recent acquisition and so the geoscientist will need to analyze acquisition reports to understand the specifics of the survey. This chapter will focus on survey technology and concepts designed for the use in the energy industry.

T. Tylor-Jones and L. Azevedo, *A Practical Guide to Seismic Reservoir Characterization*, Advances in Oil and Gas Exploration & Production, https://doi.org/10.1007/978-3-030-99854-7_3

3.1 Seismic Energy Sources

A seismic survey cannot be carried out without some type of seismic source to generate and release energy into the subsurface. A seismic source can be defined as a localized area where a sudden release of seismic energy occurs, which propagates downwards and results in stress and strain (Chap. 2) of the surrounding medium. Seismic energy can be generated by a naturally occurring source (i.e., a passive source) or an artificial source which is a man-made device (i.e., an active source).

One clear difference between these source types exists around control and the amount of energy released. An active source can be designed and controlled in a specific way to fit the needs of a survey and can also produce a repeatable source wave character whenever required. With a passive source both the source wave character and the timing cannot be controlled. As such, a survey with a passive energy source is more of a listening experiment than a defined and controlled experiment, as in the case of the active source. The most common passive source that comes to mind is an earthquake or the interaction between the ocean and land. For an active source we may think of the energy release from a man-made explosion. Passive sources can be variable and unpredictable which presents challenges when it comes to processing the recorded data. It is fair to say that both active and passive sources are useful given the right application within the right setting, and depending on the technique used the resulting data will have certain advantages and limitations of other methods.

3.2 Passive Source Survey

Passive source seismic surveys are not frequently used in the energy industry, but several successful application examples are available in the literature (Table 3.1). Surveys recording signals generated by natural sources often deal with data recorded at low-frequencies, typically between 0

and 10 Hz (Fig. 3.1). The large variety and different natural processes that passive methods exploit have kept them as a more specialist seismic survey rather than being mainstream. The energy industry has for many decades been chasing improvements to seismic resolution (Sect. 2.2.6) as its primary goal. Many of the passive methods described in Table 3.1 are useful for identifying large-scale subsurface features (>100 m), in contrast to modern active source techniques that identify both large- and small-scale features (<10 m). This is a primary reason why passive source techniques have been overlooked, yet not all geological challenges can be solved with conventional active source methods.

Recent improvements in access to high-performance computing, along with energy resources exploration and development opportunities becoming more geologically complex, has inevitably led to more research and funding for passive source techniques. These techniques can be intergrated with active sources to produce a blended seismic acquisition and to aim for higher sampling density and, consequently, data with improved resolution. Another strong driver for industry uptake of passive source methods is environmental reasons; specifically to address concerns about the dangers of large active source surveys on marine wildlife. Additionally, the energy required to power large active source surveys, whether onshore or offshore, is carbon intensive and although many operators and seismic acquisition companies plan surveys to minimize the carbon footprint, it remains a concern for the energy industry moving forward. That passive source surveys are carbon efficient may be what eventually brings them into the mainstream, where their potential can be proved.

These passive techniques might have some advantages over active source seismic methods, summarized below:

(a) These methods are orders of magnitude cheaper than acquiring a complex modern active seismic survey;

(b) A flexible solution for onshore areas with rough terrain or mountainous regions resulting in poor illumination (e.g., shadow zones)

Table 3.1 Summary of some passive seismic methods with applications in the energy industry

Method name	Description	Illustrative application examples
Ambient noise tomography	Uses ambient noise sources such as atmospheric activities or ocean waves received at temporary or permanent seismic arrays to model the S-wave structure of the subsurface	Chmiel et al. (2019)
Interferometric seismic imaging	The method is based on the principles around interferometry where a light beam can be used to sample the property of an object or medium and is then combined with a reference beam. The resulting interference pattern is called an interferogram	Hornby and Yu (2007) Curtis et al. (2006)
Earthquake tomography	This uses local seismic events such as earthquakes as the main source and the ray paths are analysed by a local network of receivers	Kapotas et al. (2003)
Hydrocarbon microtremor analysis	The method focuses on the frequencies between 1 and 6 Hz that are generated by background ambient noise and on variations seen in this frequency range above a given reservoir area	Prabowo et al. (2017)

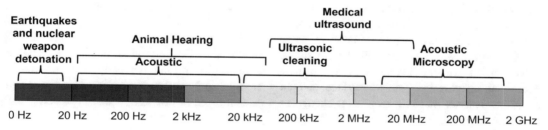

Fig. 3.1 The frequency spectrum between 0 Hz and 2 GHz and examples of its applications

due to limited access for acquisition trucks or drilling equipment;

(c) The low-frequency content is useful for constraining techniques like FWI;

(d) Environmentally sensitive for onshore and offshore areas, where either wildlife or natural habitat would be directly affected by a seismic survey;

(e) Broad data coverage for large-scale (> 1000 kms) regional understanding;

(f) Application for geotechnical characterization and engineering works where parameters such as soil strength and shear stress are important.

3.3 Active Source Survey

Active source seismic surveys are the default choice for use in the energy industry whether for exploration or development activities. The evolution of active sources has been progressing since the first seismic surveys were recorded using sticks of dynamite, and the field continues to be a hot research topic for both seismic acquisition contractors and production and operating companies. There are several fundamental factors that keep active sources as the preferred method for acoustic wave generation:

(a) The ability to control spatial deployment in a seismic survey;

(b) The control on the characteristic of the energy pulse emitted;

(c) The timing or frequency of when energy is emitted.

These three features allow for active source surveys to be highly customizable for individual projects, which increases the value of the data recorded.

The design of acquisition sources used within active sources is complex as there are many components that must be carefully planned and deployed in the field to fulfil the objectives of the study, minimizing the chances of failure. The level of complexity in active source surveys is considerably larger than for passive source surveys. Preparing a survey design and executing it within acceptable acquisition parameters for a specific objective can be the difference between producing an average data set with no real new additional insights into the subsurface and an excellent data set that is a clear step change in understanding the area of interest. Of great importance when considering the source for survey design is the ability to send acoustic energy deep enough into the subsurface to reach the target and produce recordable reflections. Not only must energy reach the target, but it must still have sufficient energy and frequency content to give clear, detailed reflections to be recorded. Planning the source configuration and how it will be moved over the survey involves many factors including water depths, subsurface geology, regional structural fabric orientation (e.g., faults, fractures), depth of zone of interest, anisotropy and subsurface attenuation structures (e.g., salt layers or bodies).

Seismic processing techniques can produce useful images even with old data or data acquired with non-optimum survey designs. However, limitations occur when the data have signal that is too weak or simply not present at the feature to be imaged. For large offshore surveys using the latest technology costing $10 s of millions, a poorly planned survey will be a poor investment. The planning should always be done using as much integration of key subsurface disciplines at the earliest possible time.

3.4 Offshore Survey

3.4.1 Marine Sources

The offshore source domain is dominated by a particular single-source type: the airgun (Fig. 3.2). Historically, explosives were a widely used marine source and required charges ranging in size from 25 to 50 kg exploded near the sea surface. This technique had a range of disadvantages including environmental concerns, the need to transport huge amounts of explosive, expense, lack of precision about the source depth, restrictions of operation and inconsistent rate of fire. Several other unusual techniques were trialled during the early days of marine surveys, such as the steam gun, which involved generating high-pressure steam in a boiler and pushing it into a valve along ridge pipes below the sea surface, before releasing it.

The water gun was perhaps the most successful air gun alternative of these early marine sources, and geoscientists may still come across vintage surveys in which it was used. Essentially an adapted airgun, but instead of releasing pressurized air into the water column, pressurized air pushed a piston to jet out pressurized water. Water guns tended to produce a complex signature and were also less repeatable than airguns when deployed in the field. By the 1970s these early techniques had all but vanished with the airgun becoming the preferred marine source for the next 50 years.

The airgun (Fig. 3.3) is an acoustic seismic source powered by compressed air under the surface of the water. The basic mechanics work by filling a chamber in a metal airgun with very high-pressure (10–15 MPa) compressed air which is fed from a compressor onboard the vessel. The compressed air is stored in one

Fig. 3.2   (Top) Single air-gun and three-gun cluster. (Bottom) Pictures of air bubbles at 1ms, 1.5ms and 7ms after firing a small airgun in a tank (LandrØ and Amundsen 2010). (Image courtesy of GEO ExPro)

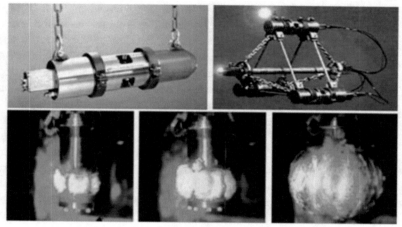

Fig. 3.3   Schematic representation of an airgun charged and after discharge

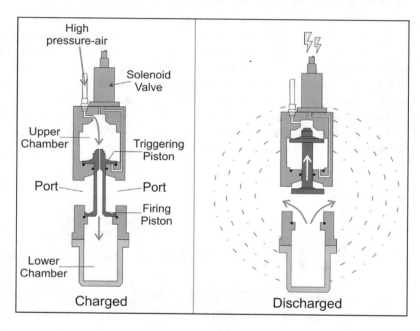

chamber and sealed by an electronically controlled sliding valve. An electrical signal activates a solenoid that opens a pathway for the compressed air to be moved in the second chamber, thus discharging the compressed air into the water column. The entire process happens in milliseconds after which the gun resets for another shot to be discharged quickly.

The ports by which the gas escapes vary in shape and size and can give the released energy specific characteristics in terms of amplitude and frequency content. Depending on the objective of

the survey it may be desirable to put more energy into specific frequency ranges.

The size of the chamber in the gun affects the energy and frequency component of the source and is, therefore, an important variable to be considered when planning a survey, processing seismic data or interpreting a seismic data set. The airgun volumes are generally measured in cubic inches representing the volume of air the chamber can hold. The amplitude of the acoustic wave produced is directly proportional to the cube root of the volume of the airgun. This means that doubling the amplitude of the acoustic pulse requires more than just doubling the gun cavity. Historically, the use of airguns centred around fewer bigger guns which translated to more energy and better subsurface imaging. This was due to improved energy penetration into the subsurface that resulted in stronger recorded reflections and an increased signal-to-noise ratio.

However, a bigger gun requires increased time to recharge so the time between shots is increased leading to lower shot density and shot overlap. Another pitfall of using a single big gun is the focus of power at the very start of the shot, which is then relatively short lived and low in amplitude. To avoid the issues of single large guns, many smaller guns, referred to as gun arrays (Fig. 3.4), are used in the energy industry today. Typically, air pressures range from 1000 to 3000 psi. The array consists of between three to six sub-arrays which are referred to as strings. Each of the individual strings contains a series of six to eight individual guns. As such, a full array being towed by a single vessel may include anywhere from 18 to 48 guns. A standard survey might have the guns positioned three to ten metres beneath the surface of the water and they are fired at regular intervals, depending on the vessel speed, water depth, the maximum distance between source and receiver and the depth of the target to be imaged.

The seismic gun array is a clever solution to several potential issues that are difficult to solve with a single gun. The gun array volume is the sum of the volume capacity of each gun, so large amounts of energy can be produced in this way as well as providing a much bigger spatial spread of the source energy. Therefore, more high-amplitude energy is propagated directly vertically downward as a sum of the array, with the ratio of non-vertical relative energy being less when compared to that from a single gun. Having guns clustered together also helps with the reduction of bubble motion and oscillations in the water. The primary pulse generated from the airgun release is then followed by a series of "bubble trains" that extend the overall length of the pulse.

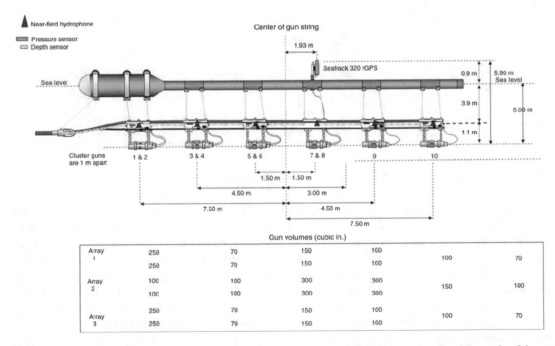

Array	1 & 2	3 & 4	5 & 6	7 & 8	9	10
Array 1	250	70	150	100	100	70
	250	70	150	100		
Array 2	100	100	300	300	150	100
	100	100	300	300		
Array 3	250	70	150	100	100	70
	250	70	150	100		

Fig. 3.4 (Top) Single air-gun and three-gun cluster. (Bottom) Pictures of air bubbles at 1, 1.5 and 7 ms after firing a small airgun in a tank (Image courtesy of GEO ExPro)

The energy pulse produced by an airgun array is commonly known as the signature and is numerically represented as a wavelet (Chap. 5). This can be artificially modelled and designed based on the gun specifications and array configuration. An effective gun array must comprise a number of standard characteristics. The strength of the gun signature and that of the actual pulse produced should be very close, which is known as the primary-to-bubble ratio (PBR) measured by the near-field hydrophone. The source pulse must produce enough acoustic pressure to be effective given the survey objectives. A measurement is made at the far-field point where the complete output of the gun array can be measured, which may be up to 300 m below the source array. This can then be used to artificially calculate the source pressure level at closer proximity (1 m) to the source array (nominal point-source level). P-P (peak to peak) strength can be measured to look at the difference between the absolute amplitude of primary and ghost arrivals at the receivers.

3.4.2   Bubbles

The behaviour of the bubble generated by an airgun is related to the pressure inside that initially exceeds the external pressure of the surrounding water column, which acts as confining pressure upon the bubble's surface. When the expanding bubble reaches a point where the internal pressure is no longer greater or even equal (i.e., equilibrium) to the hydrostatic pressure of the surrounding water column, the bubble will collapse. The collapsing bubble reaches a state where its internal pressure again exceeds the external hydrostatic pressure and so begins to expand. This cycle repeats with a period from ten to hundreds of milliseconds. The oscillation of bubbles after their initial expansion phase leads to unwanted energy signals that interfere with the recorded signal and add both low- and high-frequency reverberation into the data. This is commonly known as bubble noise (or bubble pulse) (Fig. 3.5). Large bubbles that produce low-frequency noise are most problematic for the data.

As the bubble breaks into smaller bubbles the frequency of the noise will increase, but small bubbles are much less likely to be in-phase with each other.

When the bubbles reach the sea surface and burst, they can also cause unwanted noise and interfere with the primary bubble signature. One of the main advantage of using gun arrays over single guns is the reduction of bubble noise. Gun arrays can be set up in such a way that varying gun sizes produce slightly different bubble sizes. Different bubble periods lead to destructive interference of the bubbles which helps to greatly reduce bubble noise and produce a cleaner first pulse. The dominant frequency of each bubble depends on the gun volume and, as a general rule, larger guns emit lower frequencies and smaller guns produce higher frequencies; however, the depth of the gun in the water column also plays an important part, and to a lesser part is the gun pressure.

Ideally the geoscientist wants a broad range of frequencies, both low and high, in the seismic data. The high frequencies are related to the energy and speed of release of the air from the gun. The faster the air escapes the gun chamber the higher the pressure of wave energy in the dominant first bubble, leading to more higher frequencies coming from the source. Since hydrostatic pressure varies with depths, at shallow depths the lower pressure allows bubbles to exist for a longer period with larger oscillations. Therefore, towing source arrays at shallow depths (e.g., 5m below sea surface) gives good high frequencies and resolution in the data, because the bubble time period represents the fundamental frequency, which leads to richer high frequencies. Conversely, towing at shallow depths can cause the data to suffer from sea and weather noise.

Hydrostatic pressure increases with depth due to increased water volume above, so towing deeper (e.g., 20–25m) means the bubbles exist for shorter periods and are subsequently less rich in high frequencies. However, deep towing removes the problem of surface noise and produces good low-frequencies in the data.

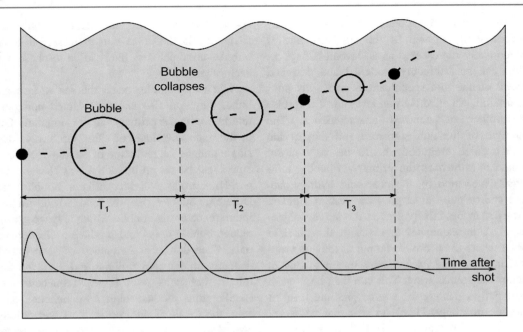

Fig. 3.5 Schematic representation of the bubble pulse effect over time

The energy coming from the bubbles travels predominantly vertically down to the seafloor. Energy is transmitted horizontally but at much lower amplitude levels. Along with the downward wave energy, part of the energy simultaneously travels upward to the sea surface before being reflected at the sea surface and travelling down to join the original downward travelling wave from the bubble. This reflected wave is referred to as a source ghost and makes up part of the source wavefield signature. The importance of the source ghost on the frequency content of the seismic data is discussed later in this chapter (Sect. 3.4.6).

3.4.3 Multi-source Surveys

The most basic source setup is towing a single gun array. For most modern surveys, the standard is towing two gun arrays firing alternatively in what is known as flip-flop source acquisition. This setup has several advantages over a single gun array as one array can recharge while the other fires, thus, reducing the time between shots and so increasing shot density. Two gun arrays also produce a better image due to an increase in sampling in the crossline direction of the survey, which is under sampled compared to the inline direction in a single-source array survey.

Adding a third gun array makes the source firing pattern flip-flop-flap and further improves the advantages of the dual source survey. The third gun array gives the option to use fewer but more widely spaced streamers, unless there is a specific requirement for high-density receiver spacing. Multi-sources can also be fired in a simultaneous mode, but this presents an additional complexity of separating the individual source signals, which is done during seismic processing by a workflow called deblending. Sometimes a dither is introduced during simultaneous shooting which is a tiny timing difference between simultaneous gun firing, in the order of a few hundred milliseconds, to help the deblending process distinguish between the different recorded gun signals.

De-blending is not a new concept but its successful application to commercial acquisitions has emerged in the last 5–10 years. The technique solves the problem of overlapping recorded shots being recorded simultaneously by

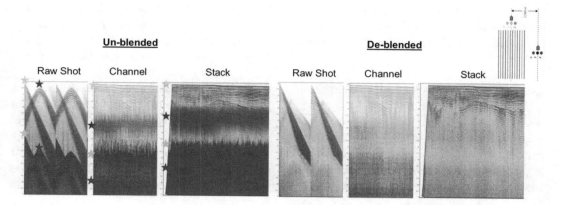

Fig. 3.6 Example of seismic data with multiple source shots present, (Right) seismic data from the same survey but with de-blending. (Images courtesy of CGG Earth Data)

separating them out into individual shots (Fig. 3.6). Complexity can come from trying to completely separate a strong amplitude signal and weak amplitude signal overlying each other without having any residual signal contaminating the wrong shot. Several methods exist, with adaptive subtraction and inversion being two widely used methods. Advances in deblending have been pushed by the demand coming from commercial acquisitions. Surveys can now be acquired with five (penta), six (hexa) and even ten (deca) sources. These emerging multi-source surveys chase finer sampling, mainly in the crossline direction at the cost of some inline sampling, greater fold and increased efficiency in survey. This approach is certainly welcomed in efforts to reduce carbon emission from these types of activities.

The traditional method of firing all the guns in a single array simultaneously has been challenged in recent years. The simultaneous method aims to reduce primary bubble reverberation along with producing a high-amplitude spike pulse. The simultaneous method has operational constraints because the whole array must be charged before all the guns can be fired, which requires maintaining high supply levels of compressed air and careful management. Ziolkowski (1984) proposed the concept of recovering a coherent source signal for a single gun array even

when the individual sub-array gun is out of synchronization by 100 ms. This concept, now referred to as "popcorn shooting" (Abma and Ross 2013), was built on to develop a non-simultaneous source acquisition method which aims to spread the gun array energy over time rather than the focused high-amplitude spike. Not only does this method simplify the management of the gun arrays from an air supply and timing perspective, but it also builds flexibility into how the arrays are deployed. Using this modern concept, survey times are reduced and energy consumption minimized. One other key advantage is the reduction in environmental impact due to reduced acoustic noise of all guns firing at once.

It is not uncommon in modern surveys to have a dedicated source vessel(s) rather than a single vessel towing both the source and receivers. Dedicated source vessels are standard for any type of survey where the receivers are not towed but placed on the seabed (e.g., ocean bottom cable (OBC), ocean bottom node (OBN)). Dedicated source vessels are also used in some configurations of wide-azimuth surveys (WAZ), where the source vessel runs along a single static line and a dedicated receiver vessel moves across the survey area varying its proximity to the source vessel. These types of acquisition geometries allow for imaging the subsurface with energy that travels along different azimuths, which is important for complex geological settings such as those associated with salt tectonics.

3.4.4 Streamers and Hydrophones

Hydrophones are used in marine surveys to record seismic waves produced by the seismic sources. Hydrophones detect energy in the form of pressure change/variations. In a conventional seismic survey, multiple hydrophones are combined within long flexible cables called streamers. The streamer cables comprise many hydrophones connected with power and data cables encased in a thick plastic protective sheath (Fig. 3.7). The cables are filled with buoyant liquid, such as oil or gel, which is less dense than water and combined with the use of weights keeps the streamer horizontally and neutrally buoyant in the water column. Slanted streamers are also a possibility (Sect. 3.4.6). The streamers are towed behind the source array and beneath the sea surface to reduce noise caused by waves and swell. These cables can be referred to as receiver arrays because in a three-dimensional

seismic survey a vessel will tow multiple streamers in a single survey. The streamers are flexible and are transported on cable drums which are placed at the back of vessels so they can be easily deployed.

Hydrophones have great sensitivity and can pick up marine acoustic signals from many miles away due to seawater being such an effective transmitter. Sound waves move much faster in water (approximately 1500 m/s) than in air (350 m/s). Water temperature also effects the speed of sound waves which move faster in warm water than cold.

Hydrophones commonly use a piezoelectric transducer to record the pressure waves, converting energy from one form to another. Piezoelectric means electricity from pressure and as such is a type of electroacoustic transducer that converts electrical charge from some form of solid material into energy. A range of materials can be used such as quartz crystals or ceramics, which have the special property of altering in shape when in the presence of an electrical field. Conversely, if the materials are exposed to strain an electrical charge can build up on the materials' external surface. It is possible to measure a

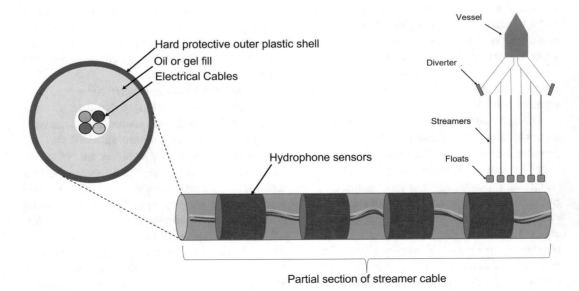

Fig. 3.7 Schematic representation of a standard streamer

voltage from this electrical field using sensors on the materials' surface. The hydrophone exposes a suitable piezoelectric material to the pressure variation from the seismic source leading to geometric distortion of the material. The corresponding electrical charge is measured as volts, which are proportional to the level of pressure required to achieve that distortion.

A near-field hydrophone (NFH) is a sensor attached to each gun in the sub-array, usually at the back or above, and is used to record pressure data at proximity to the source. The information recorded by the NFH has small offsets and can capture data that the main receiver array is unable to record. The original purpose of the NFH was for source QC, such as understanding gun timings and identifying anomalies that might affect the survey. Modern application involves using the information to improve some of the later processing stages such as de-bubbling, de-signature and de-multiple. Another interesting application of these recorded data is to provide high-resolution (~100 Hz) imaging of the shallow (0–1000 ms) subsurface. As these data can be obtained from any marine survey it may be useful for shallow hazard analysis, geotechnical characterization and shallow velocity analysis, which can then be incorporated into seismic velocity models.

There are several key elements that must be addressed when controlling the receiver setup during a marine acquisition:

(a) Controlling the depth and depth profile of the streamer;
(b) Understanding the location of each of the hydrophones relative to the source and other hydrophones;
(c) Reducing unnecessary noise that might be picked up while the receivers are recording;
(d) Gathering the data back to the vessels so it can be stored and QC'd.

Controlling the depth of streamers in the water column along with determining the position of every hydrophone is one of the major challenges of a marine survey. To solve the depth control issue, mechanical depth controllers, commonly called "birds", are positioned at intervals along the streamers. Birds adjust the angle of their fins (or wings) to correct for erratic or undesirable depth changes and can also be used to correct lateral steering adjustments caused by currents. Streamer array depths are constantly monitored during the acquisition and coarse height adjustment can be made on some vessels using an adjustable winch at the front of the streamer.

Recording the position of the streamers during a survey is done using a combination of GPS positioning and acoustic transceivers. Streamers can be subjected to strong currents during towing and may move out of alignment with the vessel. The positioning transceivers can be placed as a minimum at the front, middle and end of streamers and give information on that particular streamer. For more precise understanding of streamer positioning during acquisition, partial cross-bracing can be used where individual acoustic transceivers are placed along the streamer every 100 m. The transceivers measure their own position relative to other transceivers on the same and different streamers within a certain proximity. A further advancement is using full cross-bracing that requires transceivers along the full length of streamer; however, there are implications on cost and survey deployment complexity. Acoustic cross-bracing gives the absolute best recording of streamer position. There are also depth measurements taken by sensors and digital compasses that provide additional data to ensure the exact position of the receivers is known, which is often called the navigation data. These information are important when the survey data are processed.

Noise picked up by marine streamers can come from a range of different sources in the marine environment and even very small noises can be recorded by these sensitive hydrophones. The two most common types of noise effecting receivers are natural and man-made. Natural noise may include swell on the sea surface and noise generated as the sea interacts with the streamer as it is pulled through the water. The sea state during acquisition is important in survey planning and acquisition management. Man-made noise can be vessel noise from the survey boat or other marine activities in the vicinity of the survey. Regular noise with a distinctive character can be removed during data processing, but random variable noise is much harder to predict and consequently remove completely.

Over time the length of streamers has increased meaning more hydrophones can be put inside to record data. A short cable of only a few metres with a single hydrophone would be referred to as a single channel acquisition. The term channel refers to the interconnection between the receiver which measures the signal and the recording system which records the signal measured by the receiver(s). The more geophones the greater the requirement for multiple channels to simultaneously record all the measurements made by the hydrophones. The different receivers are connected in series and parallel to produce receiver arrays. In a modern survey, there may be hundreds or thousands of channels making streamer lengths of up to 14 km.

Older type streamers recorded analog signals which still occasionally appear in old data sets. These streamers carried out continuous recording similar to that used for monitoring earthquakes. Analog receivers required physical copper wire to transmit the analog signal as an electrical current. Streamers with large numbers of hydrophones caused cables to become thicker and heavier to house the increasing number of wires. Consequently, maintenance and possibility of cable splits and wiring issues also increased. Due to practical limitations analog streamers were restricted to several hundred channels.

In the 1960s surveys started using digital signals. The digital streamer converts the recorded analog signal to a digital signal at the geophone before sending it to the recording system. This system gathers the signal from groups of hydrophones at anytime and sends them down just a few cables to the recording system. How often this is carried out is called the sampling rate of a survey. The digital signal can be sent more efficiently and requires less wiring and so these systems can have thousands of channels. Sampling rates of towed streamer marine surveys are standardized at 2 ms or 4 ms. This is a particularly important parameter because it influences the resolution of the data and how we image a particular target or structure (Sect. 2.2.5). Modern marine surveys can carry out continuous recording by extending the receiver recording time (i.e., record length) to be longer than the time interval between consecutive shots. Recording data in this way has the advantage of reaching very deep targets as well as making the acquisition efficiency much greater.

3.4.5 Broadband Seismic Data

The term conventional, or narrowband, data is often used to refer to seismic data with a typical frequency bandwidth of 8–80 Hz. Even with what seems like a large range of frequencies in these data sets, the usable frequencies are often considerably narrower and skewed towards the low end of the bandwidth. Geoscientists continue to explore and develop both deeper geological targets and conventional depth targets with varied geological complexities. Once the domain of the specialist, the demand for complex seismic-based attributes and seismic rock property volumes is now a pre-requisite in most standard projects. Conventional seismic data often falls short in delivering all the geological and geophysical insights required for a project. Today a combination of seismic acquisition and processing techniques have evolved to produce data with broader bandwidth called broadband data (Fig 3.8). Historically, useable low (<8 Hz) and high frequencies (>60 Hz) were absent or difficult to retrieve from marine seismic data until 10–15 years ago when they became a dominant feature of the new seismic data sets offered by seismic acquisition companies. Broadband data are now recording usable frequencies as low as 2–3 Hz and high as 100 Hz. This broadband provides a number of key imaging and interpretation advantages to conventional data:

(a) Deeper investigation: Low-frequency waves have a longer wavelength able to penetrate deeper into the earth and are less effected by attenuation of complex overburden or large velocity changes seen at unconformities;

(b) Sharper detail: To define the top and base of a layer correctly, especially thin beds,

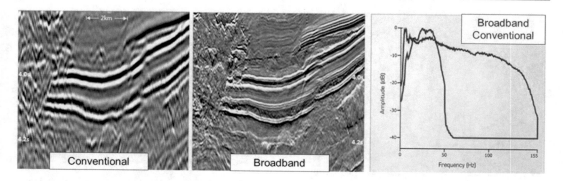

Fig. 3.8 (Left) Conventional seismic data (Middle) Modern broadband (Right) Comparison of the frequency content. (Images courtesy of CGG Earth Data)

requires both high and low-frequencies. Broadband data allows sharper definition of beds;

(c) Reduction of wavelet sidelobe: Conventional reflectivity data contains sidelobe energy from the wavelet that can lead to interpretation mistakes and hides true geology. Increased low and high frequencies greatly reduce sidelobe events in the wavelets and narrow the energy spike while increasing its energy, resulting in reflections relating to real geological events;

(d) Seismic inversion: Absolute seismic inversions require the missing seismic low-frequencies between 0 and 8 Hz and often rely on overly simplified low-frequency models (LFM) based on wells to fill this gap. Broadband data reduces the reliance on these models, with only the lowest frequencies (0–3 Hz) relying on LFM (Fig. 3.9).

3.4.6 Ghost Reflections

Broadband data are associated with the removal of the seismic ghost reflection which is generated at the source array (source ghost) and the receiver array (receiver ghost). The ghost is responsible for the creation of notches in the frequency spectrum of the data that damages the ability to recover both low and high frequencies. Research about ghosts in marine seismic data is not a new

topic; in fact, this phenomenon was described as early as 1953 by Van Melle and Weatherburn.

When a marine airgun produces a pressure pulse in the vertical direction energy travels both downward and upward simultaneously with the upward waves going to the sea surface (Fig. 3.10). This energy is then reflected from the sea surface where, due to its interaction with a large enough velocity contrast (i.e., water to air), it flips polarity before continuing downward with the rest of the down-going energy. This source ghost is considered to be a combined part of the source wavefield. The ghost cannot be simply distinguished from the primary wave by a difference in its arrival time at the receiver because they arrive at virtually the same time. The source ghost is problematic when it reaches the receiver and manifests itself as periodic attenuation "notches" in the seismic frequency. Figure 3.10 shows the different types of ghosts that can appear in seismic data. Here, in the frequency spectrum, the notches appear to be dependent on the depth of the source and receivers beneath the sea surface.

To optimize the removal of the receiver ghosts, a conventional source can be combined with a broadband streamer configuration and de-ghosting processing. Some broadband acquisitions tow the streamers with a slanted or curved profile which diversifies the notches so that they occur at different frequencies on different channels. Then multi-channel processing can recover the data over a wide spectrum. Another

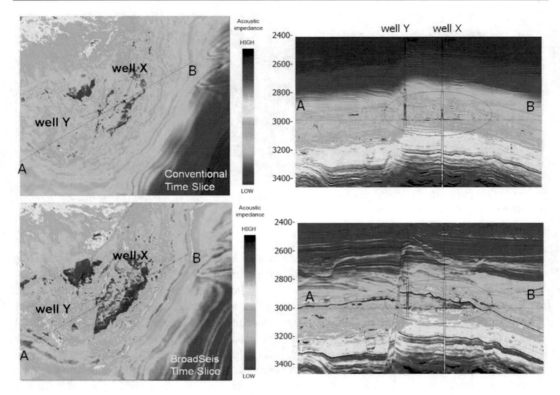

Fig. 3.9 (Top) Post-stack acoustic inversion of datafrom Brazil's Santos Basin, (Bottom) Post-stack acoustic inversion using broadband seismic data showing greater dynamic range and more realistic correlation with geology and better match to blind well. (Images courtesy of CGG)

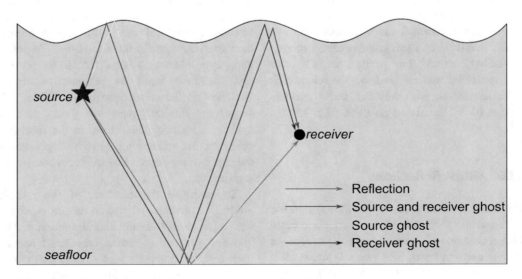

Fig. 3.10 Schematic representation of a source and receiver ghost for a seismic survey

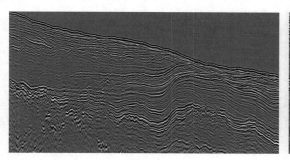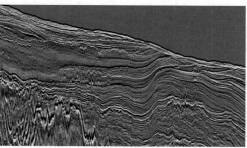

Fig. 3.11 (Left) Seismic data with no removal of ghost notches, (Right) Fully de-ghosted (source and receiver) seismic with extended bandwidth and healed ghost notches. (Seismic data images courtesy of TGS)

mainstream de-ghosting method favours using a combination of dual hydrophone and velocity sensors in the streamers, to show the direction of the wave being recorded. The recorded receiver ghost can, therefore, easily be identified and removed via summation of both the hydrophone and the vertical velocity data.

Broadband source solutions goes further in an attempt to attenuate the source ghost through positioning the source guns at different depths in the water column. The concept works by breaking up the ability for the upgoing energy (cause of the ghost) to be constructive while allowing the down-going energy to still build up constructively. Guns are fired sequentially to produce variable signals which allow the different ghost signatures to be identified and subsequently removed during processing. Essentially, this method attempts to delay the part of the wavefield that causes the ghost reflection. This source solution is not without its drawbacks. This mechanism of ghost attenuation can have a negative effect on the energy of the down-going waves if the gun fire is too closely synchronized which reduces the ratio of peak amplitude of the initial pulse with the amplitude of the bubble oscillation. The deeper the sources are in the water column the greater the confining pressure being placed on the bubbles, which results in the reduction of bubble size and, as a direct consequence, a reduction in high-frequency content. As well as focusing on an acquisition solution for the source ghost, modern surveys look at multi-level sources to reduce its impact. De-ghosting can also be done with a purely processing based

approach which utilizes factors like the dimensions of the survey, the position of the sea surface and tide data along with survey geometry to estimate a measurement of the ghost characteristic and timing, which can then be removed from the recorded receiver data.

Modern processing solutions can attenuate effectively both the source and receiver ghosts. Once the data has been de-ghosted and the ghost notches are removed, the frequency spectrum can be shaped to broaden the frequency bandwidth for both high and low-frequencies. This step is often applied too cautiously so best practice is to try several operators until the desired bandwidth is reached within the limits of the data. The de-ghosting processing workflow is highly recommended for re-processing vintage marine seismic data sets. The results of careful de-ghosting can transform a vintage data set into appearing almost modern with much richer low and high frequencies (Fig. 3.11).

3.4.7 Survey Geometries

A marine three-dimensional seismic survey requires a great deal of acquisition equipment that fits together, to produce the desired seismic data. How the equipment is arranged (the distances between equipment and specific surface depths) directly effect how the data will be recorded. This is referred to as the acquisition geometry and it has an effect on key parameters such as resolution and fold of the seismic data. When working with a new seismic data set it is

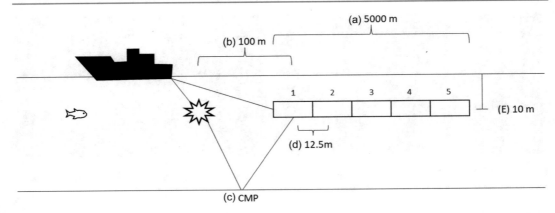

Fig. 3.12 A typical seismic survey geometry with **a** total streamer length of 5 km, **b** near-offset distance, **c** represents the reflection point happening at half-distance between the shot and the first hydrophone **d** receiver spacing at 12.5 m and **e** depth of source at 10 m

important for the interpreter to have access to the acquisition report and relevant processing report which will detail the geometry and processing parameters used. It is even more important to understand the differences in acquisition setup and geometry when working with legacy data or a range of different survey vintages. Knowing the variation in streamer length, receiver spacing and even gun size may help explain why data character changes over an area and how much caution should be applied to variations in amplitude, phase and frequency.

Figure 3.12 illustrates a simplified image of the main components of a seismic streamer setup. For a streamer survey, the common variables will be the depth of the source; distance between the source and the first hydrophone (i.e., the near-offset); spacing between hydrophone and total length of the hydrophone cable (i.e., streamer). A description of each element is provided below:

(a) Streamer length: Important for understanding the depth of recorded data and offset/angle data ranges;

(b) Near offset: Distance from the shot to the first receiver group;

(c) Common mid-point (CMP): The mid-point between the shot and the receiver. In Fig. 3.12, it is illustrated for the first recording group;

(d) The group spacing defines cable geometry along with the number of groups, which can be in the hundreds;

(e) Streamer depth is one of the most important measurements for the correction of the data to the datum of mean sea level and also calculating the ghost notch frequencies.

Figure 3.13 shows the simple geometry typically expected in a modern marine three-dimensional seismic survey, which contains multiple sources and tow several long (8–14 km) streamer arrays behind the boat at a single time.

3.4.7.1 Shot Point Interval

The frequency with which a gun array fires between two shots is referred to as the shot point interval. Shot point interval varies depending on the survey design along with the number and size of the source array, required density of the survey data and the vessel speed. The recording from a single shot is called a shot record. If a vessel is travelling at a speed of 5 knots, equal to 2.5 m/s, then a ten-second delay between a single gun array firing would put the shot spacing at 25 m. Traditionally, surveys with multiple source arrays are fired alternatively giving time for the other gun array to repressurize ready for the next shot. However, we discussed that modern multi-source gun timings are becoming less standard (e.g., triple, quad and penta shooting)

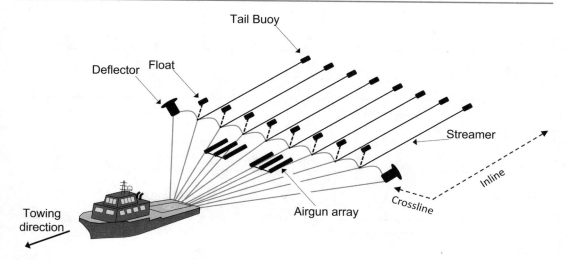

Fig. 3.13   A standard marine three-dimensional geometry running multiple source arrays and multiple streamers

(Sect. 3.4.3) as deblending techniques become ever more advanced.

3.4.7.2   Fold, CMP and Binning

Seismic fold is a measure of redundancy within the seismic survey. As the vessel moves, the shots are fired and the reflections recorded by the receivers, but if the distance moved by the vessel is quite short then the energy from the next shot is likely to produce reflections from some of the same points in the subsurface, but with slightly different ray paths. We are, in effect, getting continuous coverage. To calculate the fold, we need to know the speed of the vessel, the shot interval and the group interval spacing. Figure 3.14 shows an example of how seismic fold is build up.

For any survey, the geoscientist wants as high fold as possible to show that the survey has produced a lot of reflections from the subsurface over a small area. This concept can be thought as the density of spatial sampling. The survey fold is more variable at the edge of the survey because of decreased shot density. Therefore, it is good practice to review a seismic fold map, if available, for any survey. In an area of reduced fold, the processed image may look different simply due to less dense spatial sampling, which can be mistakenly interpreted as an indicator of lithology variation and even variation in pore fluid fill (e.g., presence of hydrocarbons).

Common mid-point shooting (CMP) is the term used to describe acquisition geometries that produce collections of reflections from the same subsurface point. The term common depth-point (CDP) is often used interchangeably with CMP, however this can only be used for the flat laterally invariant reflector. Before seismic processing begins, the spatial position of a reflector is unknown so a useful approximation is made during processing somewhere vertically down mid-way between the source and receiver groups. The application of CMP is useful for flat geology and when there are no lateral variations in velocity as it follows some simple geometric laws, but when we introduce dipping geology with lateral varying velocities the CMP method is no longer positioned at the true mid-point.

The area that defines a CMP location is called a CMP bin. Bins can be thought of as a small, defined equally-sized spatial area within a three-dimensional survey that is used to organize and categorize recorded reflections based on their location. Traces with mid-points that fall into a particular bin are grouped together. The number of traces in a bin is what defines the fold of the survey.

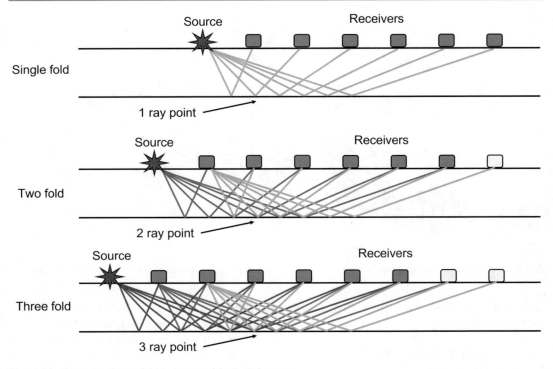

Fig. 3.14 Example of how fold is increased in a marine survey

Although simple to describe the complexities of CMP calculations are huge for a large three-dimensional survey. From a single shot for a single receiver line, 150 mid-points can be generated which is then multiplied by multi-sources and receivers and all the shots in the survey. Additional complexity comes from converted P-S waves (Fig. 3.15). The problem is solved using a combination of the survey geometry components and ray paths.

Binning is an important part of seismic processing as the size of the bin used can vary the effectiveness of signal-to-noise enhancement during CMP stacking, and can also affect the strength of reflection events observed in the final processed image. These are represented as CMP traces in a seismogram. Figure 3.16 shows an example of how a CMP point is defined as halfway between the shot point and receiver position.

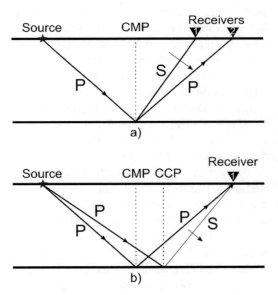

Fig. 3.15 a Conversion of an incident P-wave to a reflecting S-wave, **b** P-P and P-S reflections

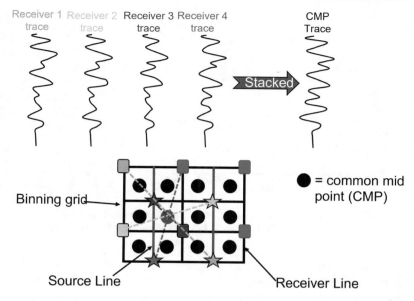

Fig. 3.16 How traces from different sources are organized using CMP and binning (adapted from Chaouch and Mari 2006)

3.4.8 OBN and OBC

Ocean bottom seismic (OBS) is a name given to seismic data recorded from hydrophones placed on the ocean floor rather than from receivers towed in the water column. Unlike marine streamers, which only sense or record pressure changes in the water column using hydrophones, ocean floor receivers record movement of the sea floor and, like receivers used for onshore surveys, they are equipped with geophones as well as hydrophones.

There are two OBS survey methods: ocean bottom cable (OBC) and ocean bottom node (OBN). These two types of survey techniques should be considered more advanced than a streamer survey. They are certainly more expensive and require more equipment as well as more complex data processing techniques. This increased effort provides the geoscientist with a range of advantages:

(a) Easier receiver station repeatability for time-lapse (four-dimensional) seismic data;
(b) The ability to record shear waves;
(c) Flexibility to shoot around infrastructure (e.g., platforms, offshore wind generators) without the need for complex undershooting;

(d) Removes the need for streamer steering solution and mitigation against currents;
(e) Greatly minimizes noise associated with the sea surface;
(f) Ideally designed to record multi-azimuthal data;
(g) Repeatable system for life of field seismic monitoring.

OBC

The concept of OBC for seismic surveys was actually conceptualized as early as 1938 (Ewing and Vine 1938). Through the 1950s OBC became used as an extension of onshore surveys going into shallow water to around 10 m, when they were known as bay cable systems (Hovland 2016). OBC field tests were carried out in the Gulf of Mexico in the early 1980s to address the problems of strong currents, navigational obstacles and towing noise associated with streamer surveys (Zachariadis et al. 1983). In 1998 results were published from the UK North Sea Alba field of the first OBC (two-dimensional field trial and full three-dimensional survey) survey applications that specifically made use of shear waves' response to the presence of oil, which was not fully imaged by P-waves (Macleod et al. 1999) (Fig. 3.17).

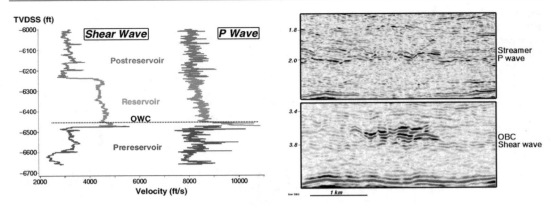

Fig. 3.17 (left) Dipole sonic log through the Alba reservoir showing sensitivity of S-waves to oil. (right) Comparison of streamer data P-wave (1989) and OBC P-S data at the Alba Reservoir (MacLeod et al. 1999)

A typical OBC receiver array comprises vertically orientated geophones and hydrophones wired together like a streamer, which can be temporarily laid or permanently anchored to the seafloor (Fig. 3.18). Initially these systems were expensive and time-consuming and so designed for use on large fields that could justify the expenditure of such a seismic acquisition solution. Advantages were focused mainly on ease of acquisition around existing infrastructure such as platforms and large subsea infrastructure that were too high risk to shoot directly over with streamers. Modern OBC surveys will contain multi-component (4C) receivers able to record not only P-waves but also S-waves.

The advantage of recording both P- and S-waves is shown in the Alba example and has been applied in other areas for challenges like imaging through attenuation zones such as gas clouds. Another clear advantage of having the receivers in a fixed position is for the application of four-dimensional time-lapse shooting over the life of a field. The concept of four-dimensional acquisition relies on an initial baseline survey being shot prior to any or little field production, with subsequent surveys being shot to find

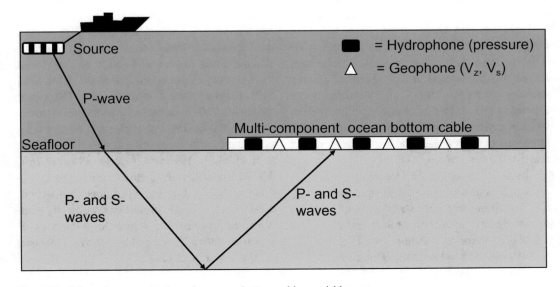

Fig. 3.18 Schematic representation of an ocean bottom cable acquisition setup

differences in the seismic response as the fluids and rock properties in the reservoir change in reaction to the effect of production. This technique is also very effective at monitoring sweep efficiency and identifying unswept pay where horizonal injection wells are used to increase recovery and carbon geo-storage monitoring. One of the biggest advantages is the possibility to acquire multi-azimuth data, which have far superior illumination of a feature or target from multiple directions to that of a single towing direction. Consequently, high quality imaging of geological features like salt structures and fault complexes, not seen from single azimuth streamer data, can be interpreted.

The peak success of OBC has been relatively short lived as streamer surveys offered broadband solutions in the early 2000s at a fraction of the price. Additionally, improvements in navigation and positioning systems made streamer surveys more viable options for four-dimensional surveys, which is one of the greatest strengths of OBC technology. Deployment and initialization of OBC is a time-consuming and complex task requiring specialist boat crews, as well as accessibility issues around maintenance. Finally, the size of fields discovered today tend to be smaller and the length of production life much shorter, which may have also played a part in the decrease of OBC surveys. Although still a relevant and powerful technology its market share is much smaller compared to that of streamer and OBN.

OBN

Like the OBC method the concept of OBN began several years before its widely accepted commercial application. During the early 1990s tests were carried in the North Sea using a node-based seismic system. Today the technology is an embedded part of the mainstream seismic solutions available to the energy industry.

OBN also utilizes receivers on the sea floor but in a slightly different way to OBC. Seismic receiver nodes, which are multi-component seismometers, are independently placed on the seafloor based on the survey design. Of the two main methods for this process, the first uses a remotely operated vehicle (ROV) to accurately place the

nodes in a pattern on the sea floor. Modern ROV solutions can carry tens of nodes on a single ROV and reload from a central submerged node hopper, which keeps the whole deployment mostly subsea (Fig. 3.19). The most efficient systems can deliver deployments of up to 200 nodes in a 24 h period. Once the source has been towed over the receivers and the receivers have finished recording, they are retrieved by the ROV and brought to the surface. The receivers record continually while on the seafloor which allows for a huge amount of information to be gathered.

The second method involves stringing the nodes onto a cable at the correct spacing prior to submersion. This technique is often referred to as "nodes on a rope". A boat tows the cable dropping the nodes behind which then sink through the water column to the seafloor. Ideally, these devices are deployed along a straight line but often underwater currents do not allow for this. Once recording has finished the cable is floated to the surface via a radio-controlled buoy which is then recovered along with the nodes. This method is an overall faster than the ROV option.

OBN has eclipsed OBC as the preferred OBS method of choice. The data quality is comparable given the same vintage, but OBN has the advantage of very flexible receiver placement and being much easier to deploy. As such, OBN has moved away from being a tool only used for large developments with CAPEX budgets able to absorb the initial installation cost, timing and ongoing maintenance, and has now found a place in areas of appraisal and even exploration. Although OBN is still expensive compared to streamer surveys, which are currently the workhorse of the offshore seismic survey Industry, they are increasingly being considered as a valid alternative (Fig. 3.20). The industry is now seeing larger scale (1000 km^2) regional surveys being acquired with node systems, which indicates that the price and effort required is not prohibitive for the quality of data obtained.

The field of nodal acquisition research is still very active, particularly regarding problems associated with cost and time by focusing on the deployment of nodes. Automation via smart

Fig. 3.19 The acquisition of a OBN survey using ROV node deployment (Li et al. 2019)

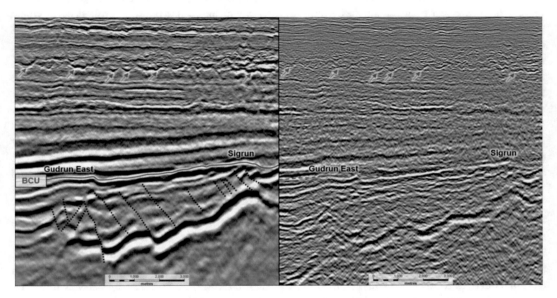

Fig. 3.20 Comparison of the difference in image quality between a 2004 3D streamer survey (Right) and a 2018 OBN survey data set (Left) over the same area in the south Viking Graben, offshore Norway. (Seismic data images courtesy of TGS)

robotic nodes with their own propulsion is a key area of ongoing research. Large groups of nodes that communicate with each other and self-deploy from position to position, based on the survey design, would greatly speed up acquisition and remove human error and risk (Brown 2011).

3.5   Onshore Sources

This section focuses on the most commonly used of onshore sources used for seismic surveys. The techniques can be broadly separated by how the source waves are produced. Some techniques use the force of gravity to accelerate a mass to either impact the ground surface at high-velocity, or use the movement of a mass to generate acoustic vibrations which are transmitted into the subsurface via a contact point, usually a metal plate. Other techniques produce pressure waves through the rapid release/transfer of energy that propagates through the subsurface.

3.5.1   Gravity Acceleration Sources

Weight drop can be considered the simplest method for creating a seismic source pulse for an onshore survey. Essentially a weight is dropped from above ground level (from 1 to 5 m) directly onto the ground surface or a base plate, where the signal is measured upon impact. Lifting the weight above ground level gives it potential gravitational energy which when dropped falls at speed due to acceleration by gravity. The weight gains kinetic energy converted from the potential gravitational energy at the start of the fall. When the object hits the ground, the kinetic energy is transferred into the subsurface. If the collision is elastic some of the energy will be lost as a sound wave as the object impacts and some will be lost in deforming/fracturing the ground surface. However, some energy will be transferred into the ground and will propagate as a seismic wave. Small trucks carry the source setup to each source location, but in very remote and difficult terrain helicopters can be used to drop weights onto the ground. The drawbacks of this method are the logistics of transporting and resetting the elevated position of the weight for every shot. Additionally, the repeatability of the source signature (i.e., amplitude and frequency) and signal-to-noise is poor. It is therefore more challenging to produce as good final seismic image compared to other methods.

Accelerated weight drop uses a piston with dampers mounted on a truck. The piston accelerates a weighted hammer from a height onto a plate which is in contact with the ground surface. One of the advantages of this method over the simple weight drop is the increased repeatability of the source signature generated. Although frequency ranges produced by both weights drop and accelerated weight drop can be quite good (5–80 Hz ranges) both techniques lack the ability to generate lower frequencies below 6–8 Hz. The amplitude and depth of penetration can also be a limiting factor unless a substantially heavy weight is used.

3.5.2   Explosive Sources

In the past the use of explosives was favoured as a seismic source in both offshore and onshore surveys; however, this technique is no longer used in offshore surveys for the obvious risk and damage to the marine wildlife and habitat. Explosives are still used in onshore surveys in places with limited access or difficult terrain which prevents the use of source vehicles. The acoustic characteristic of explosives is particularly suited as a seismic source because it produces a sharp high-amplitude and wide-frequency bandwidth source signature. This is achieved by rapidly expanding gas generating a pressure wave which exerts sudden pressure in the surrounding air and ground. Explosive sources are also referred to as pulse sources because they emit a very short burst of energy. Explosive charges are typically placed inside drilled holes to prevent the pressure wave being lost in the air and instead focus it into the subsurface.

Seismic efficiency of an explosion is defined as a ratio of the radiated seismic energy compared with the total energy expelled by the explosion (Stroujkova et al. 2015). Understanding the yield and energy transfer is important to calculate the likely energy loss, which is affected by two main factors. First is the energy density and the duration of the explosion. The second is related to the rock physics properties for the surrounding geology and its ability to withstand the force of the explosion (i.e., strength), as well as how well the pressure waves propagate (i.e., compressibility) (Stroujkova et al. 2015). Fracturing is also a factor when using explosives as a seismic source. Testing under controlled conditions has shown that much of an explosion's energy can be expended on fracture propagation with only about 3% being used as radiated acoustic emissions. The rest of the energy dissipates through plastic deformation (Doll 1984; Stroujkova et al. 2015).

3.5.3   Vibration Trucks

Vibroseis is a commonly used onshore technique and is a more modern method to explosives or weight drop techniques. This type of method is preferred for large onshore acquisitions in remote and hostile environments such as deserts or Arctic tundra. The technique was first introduced in the mid-1950s by Conoco with the name Vibroseis being trademarked by the Continental Oil Company in 1953 (Meunier 2011). The basic principle of a land vibrator is illustrated in Fig. 3.21. It comprises a large mass typically a weight in the centre of the truck. An actuator generates an oscillatory force between a metal base plate which is in contact with the ground and the reaction mass, which results in the ability to generate a maximum peak force of 40,000–90,000 lbs.

The biggest difference between an explosive source and vibrator is that an explosive source creates a very fast release of energy that produces a near-perfect single spike response. Seismic vibrators, on the other hand, emit energy in what is known as a sweep signal which typically lasts twenty to thirty seconds in duration. The sweep signal changes frequency in a linear or non-linear pattern over the duration of the sweep producing a range of frequencies. Up sweep is a term used to describe frequencies that increase with time and down sweep being frequencies that start high and decrease throughout the sweep.

The frequencies used in a sweep and the speed at which the sweep is generated are restricted by the mechanical capabilities of the vibrator truck. If a sweep is too slow the trucks may find it difficult to produce the sweep consistently without variation. In contrast if the sweep is too quick

Fig. 3.21 Illustration of a Vibroseis truck

the vibrator may struggle mechanically to vibrate the mass quickly enough. Common sweep characteristics are likely to fall in a range of 2–90 Hz at a duration from 10 to 20 s. Going beyond 90 Hz may not be useful at ground surface as the short wavelength of these frequencies will lead to attenuation of the signal resulting in reduced penetration into the subsurface. Furthermore, higher frequencies and longer sweep duration will result in extended acquisition timeframe and cost.

Vibrators have been trialed in the marine setting but were found to be inferior to airguns, which were already well understood and distributed within the industry. In terms of efficiency, the amount of energy generated by a vibrator is much lower than that of an airgun array. To generate low-frequencies with a marine vibrator, a large area is required which makes towing such equipment problematic. However, as an environmentally responsible source for the marine setting the vibrator is superior to the airgun and so this type of equipment is still under development in the energy industry.

Despite its advantages, such as acquisition speed, environmental consideration, source controllability and tolerance to hostile weather conditions, the Vibroseis survey does suffer from being unable to effectively propagate seismic waves through the weathered zone. The weathered zone is a shallow zone beginning at ground level which is affected by surface events such as weather, fluvial activity, flooding systems and human activities. The seismically defined weathered layer is characterized by a low-velocity because of its unconsolidated nature and high percentage of air-filled voids. The base of this layer can coincide with the water table or beginning of bedrock. This layer can be corrected for in the seismic processing workflow using a static correction which compensates for the time delays. Whereas explosives are placed into a borehole below the weathered zone, the Vibroseis method struggles with wave propagation through this layer.

An increasingly desirable factor in acquisition methods is repeatability, which is a strong advantage that the Vibroseis method has over explosive sources. Even when using a single manufacturer and set weight/yield of explosives the subsurface can become fractured and damaged meaning it is not possible to reuse a borehole and therefore repeat the shot process exactly. Modern GPS enables the vibrator to return to the exact position. The consistency of shot amplitude and frequency range of Vibroseis also helps hugely to apply consistent seismic processing to get a final image.

Low-frequency performance of seismic vibrators has become an area of interest in recent years, perhaps driven by advances in marine broadband seismic acquisition. The availability of low-frequencies proves helpful for the inversion of seismic traces, reduces sidelobes and gives a more geological appearance to seismic data. Most conventional seismic vibrators are unable to produce stable and consistent frequencies below 5 Hz.

Vibrator trucks are used for P-wave generation, but also produce S-waves at the same time. The generated S-waves are at non-vertical angles and it can be difficult to control their direction of propagation into the subsurface. Several methods have been tested in the field to prioritize horizontal movement of the vibrator for S-wave generation rather than vertical movement. One particular method rotates the ground plate perpendicular to the X–Z-plane and deploys two vibrator trucks in parallel to vibrate 180° out of phase, which causes destructive interference for the P-waves leaving only the horizonal displacement (Edelmann 1981). Although these techniques do produce some success with S-waves, the data quality is not as good as P-wave data for imaging and is more complex to process. Therefore, P-waves tend to be the primary wave types of the Vibroseis source.

3.6 Onshore Receivers

3.6.1 Geophones

Onshore seismic waves are recorded using geophones. The term geophone comes from the Greek word *geo* meaning earth and *phone* meaning sound. Unlike the offshore environment where the sea water column and sea conditions pose challenging obstacles for receiver deployment onshore surveys tend to be more straightforward.

Since the 1930s passive analog geophone devices have used a coil and magnet to detect a signal from the seismic wave. This type of geophone is still commonly used today but over time has become smaller, lighter and tougher with increased sensitivity and improvements in distortion reduction. These analog geophones use a transducer which is designed to measure ground vibrations and convert them into an electrical signal. The transducer must be sensitive to the direction of ground motion relating to the waves that are being recorded. The distance of the sensor from the seismic event and the magnitude of the event must be considered when selecting the geophone. For example, an earthquake or nuclear explosion can have a huge magnitude allowing sensors all over the world to detect the generated seismic waves, but a small, localized event may be harder to detect and discriminate from background noise of other acoustic events. Geophones need to be calibrated and capable of recording the amplitude and frequency ranges expected and ideally minimize any signal distortion.

A seismic transducer is the mechanism inside an analog geophone which records the ground vibration using a spring-mass-damper element and displacement transducer (Fig. 3.22). As the ground oscillates due to the wave propagation, it can be measured as ground displacement, velocity or acceleration of the ground particles. The ratio of mass and spring constant determine the natural period of the system. The mechanical or electrical damping/resistance will change both the sensitivity as well as the shape of the overall transfer function of the system. The electrical signal is given in volts with the magnitude of the output produced by the transducer requiring a conversion mechanism and scaler to be applied. The output voltage is directly proportional to the velocity of the coil movement within the case which is then scaled into volts per m/s. The sensitivity of the transducer is related to the loop

within the coil and the strength of the magnetic field from the magnet. The voltage can then be converted to a physical unit such as pressure, which is expressed in Pascals.

Geophones used in reflection seismic surveys generally record reflected waves travelling vertically from the subsurface. This means that the primary interest is recording vertical motion of the Earth. Not all waves travel upward and for the onshore surveys a strong horizontal wave known as ground roll can generate a strong vertical motion that can drown out the weaker vertical signal in the geophone recording. Ground roll is usually dominated by Raleigh waves which travel along or near to the ground surface. Raleigh waves are dispersive in nature but have a velocity of less than the S-wave velocity and so are limited to below ~30 Hz. If the near-surface layer is very heterogeneous and variable it can lead to the ground roll waves behaving very randomly affecting the seismic wave dispersion paths and velocities. Ground roll is usually dealt with in the processing stage, however, it can be very difficult to remove entirely. One approach uses a geophone that records ground movement in more than just the vertical direction; the surface wave can be identified and removed because of its elliptical propagation. This process is called polarity filtering but comes with its own challenges.

Three-component geophones (3-C) have sensors that record particle velocity in three

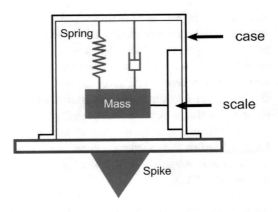

Fig. 3.22 Diagram of a transducer used in analog geophones

mutually orthogonal directions. This is especially useful in helping to determine the type of wave being recorded and its direction of approach to the receiver. The three components are a single vertical and two horizontals. When deploying 3-C receivers it is important they are correctly orientated so that the horizontal components are consistent between receivers. This can be estimated by comparing shot and receiver's azimuth based on the survey geometry and the azimuth calculated from the two horizonal components. To avoid the challenges of correct orientation of the geophone components an alternative is found with the Galperin configuration, where sensors are positioned orthogonally to each other but tilted at 54.7° to the vertical axis (Grazier 2009).

3.6.2 Digital Geophones

A recent development in geophone technology came with the introduction of micro-electro-mechanical system (MEMS). As the energy industry requires small, light onshore receivers with much wider broadband capabilities the digital geophone became the solution. Although called a digital geophone the correct term is an accelerometer, whereas analog geophones are velocimeters. The output of these geophones is digital because rather than the movement of a magnet and spring generating a voltage a change in capacitance is used. A tiny mass is attached to the spring, like in the analog geophone, for a specific direction of movement. The mass has arms that are positioned between fixed plates. When the mass moves the capacitance between the mass and the plate changes. The change in capacitance is measured and will correspond to a particular magnitude of acceleration value. Strictly speaking the MEMS sensor is based on an analog geophone but the control loop and output is from an integrated microchip and is, therefore, digital (Mougenot 2004). Gyroscopes can also be included in digital receivers and are useful when multiple components are being recorded and where placement of the receiver is not 100% controllable.

MEMS sensors have advantages such as a broad linear amplitude and constant phase response that extends from zero frequency (DC) up to frequencies above 500 Hz. They also have an extremely high resonant frequency (1 kHz), well above the seismic band. In addition, MEMS are much smaller and lighter than the transducer of an analog geophone. The dynamic range on the MEMS sensor is controlled digitally and this can have benefits for controlling the noise floor, making MEMS suited to high-resolution surveys. Besides, with no coil electromagnetic noise is not an issue. The lower noise floor however is the main disadvantage of MEMS over analog because it means any signal higher than the noise floor will be seen on all frequencies. Another disadvantage of MEMS compared to analog geophones is that they consume power and as a consequence power consumption becomes a major technical hurdle in a survey with large numbers (millions) of channels.

3.6.3 Frequencies

For the use in seismic exploration, geophones can detect particle velocity ground motion in a range of 0.000025–25 mm/s. As mentioned previously, the geophone is an oscillatory system. These systems resonate at their own frequency, if undamped, known as the resonant or natural frequency. The resonant frequencies represent the low-frequency limit of the reliable seismic data. The lower corner response is generally manufactured to a range of between 5 and 15 Hz. Frequencies below the resonant frequency can be recovered using a low-frequency transfer function. Higher frequencies in the range of 250 Hz are called spurious frequencies.

Damping is an adjustment made to the geophone using a resistor which is inserted parallel to the coil and the level of damping is expressed in percentage. The resonant frequency is dependent on the components of the geophone such as the spring, size of mass and the damping mechanism. This is important to be aware of because it determines the filter characteristic of the geophone and its measurement. If the damping factor is set to zero, the frequency of the peak amplitude will be the natural frequency at which the system oscillates. This natural frequency will produce the amplitude value for the highest peak for a given frequency. As a consequence, the geophone response goes through a large peak for the natural frequency (e.g., 10 Hz) that dominates the sensitivity, which decays away for high frequencies. At a damping of 70%, which is most commonly used, the response will be constant above the natural frequency and will define the sensitivity of the geophone. As a rule of thumb, the more damping applied to the geophone the less sensitive it will be (Meunier 2011).

The voltage response is recorded by the geophone after which it passes through several processes such as pre-amplification and low-cut/high-cut filters. The response is then converted from an analog signal into digital information which is recorded in a readable format (Fig. 3.23).

3.7 Survey Geometries

Onshore surveys can be less costly than offshore surveys mainly because the equipment is more accessible and straightforward to deploy. However, the onshore environment does come with its own set of challenges that impact the acquisition costs, such as highly variable elevations of terrain, human settlements and restrictions based on natural habitat. All these factors play a part in the design of onshore survey geometries.

3.7.1 Receiver Lines

For onshore surveys the receiver lines comprise both geophones and the associated cable required to transmit the recorded field signal back to the recording station. Groups of geophones are usually laid in defined areas known as swaths which are shot over by the source trucks before the lines are deployed in another area. Initially, the survey area is assessed and a line-clearing phase may be required to remove or flatten certain obstructions in the survey area. During the design phase desktop studies and visits to the acquisition site will be carried out to ensure that environmentally sensitive habitats remain undisturbed and disruption to human habitations or property is minimized. In the surveying phase, each of the receiver and shot positions are defined along with the correct spacing. Once the receiver positions are marked/staked out the receiver crews lay out the equipment. Sometimes receiver and shot positions are not staked for reasons such as unsuitable ground conditions or to enable quicker clean-up of a survey area; it can also speed up the preparation of the acquisition area.

The receiver cables are positioned in parallel lines, with a single swath commonly comprising six to ten cables (Fig. 3.24). The cable spacing depends on specifications from the survey design and the shooting method. Higher source density surveys employ more closely spaced lines and give greater survey data overlap. The closer the lines are together the more lines will be required to cover the survey area, which mean more receivers in the field. Issues with cable connections failing or intermittent connections is a common problem. For imaging deep targets the lines need to be longer (up to 8–10 km) to capture the longer offsets.

One of the main challenges with modern three-dimensional onshore surveys are the acquisition logistics. Deploying, collecting and maintaining hundreds of kilometres of cables and thousands of geophones, in addition to the recovery of receiver lines can be time-consuming. Consequently, geophones are constantly evolving and becoming easier to deploy and

Fig. 3.23 The process of analog geophone signal recording

Fig. 3.24 The basic geometries of an onshore three-dimensional survey

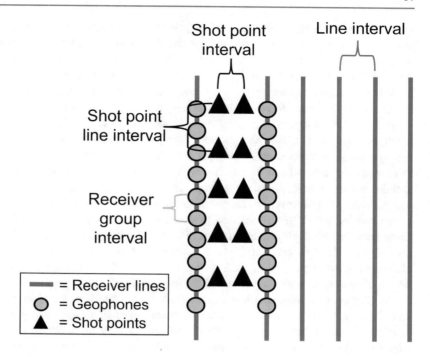

record. The recent development of a cableless system may find its application in large surveys and in areas where it is not feasible to lay cables (e.g., near actively populated areas). Though the increased weight of the geophones and their battery management brings its own separate challenges which differ depending on the survey size and environment.

The recording system for seismic data of an onshore survey is usually a mobile vehicle which performs not only the collection of the field data, but also manages the survey to ensure acquisition parameters are within the survey tolerances for raw field data. For a large three-dimensional survey with small receiver intervals, long receiver lines and dense line spacing, the number of channels required can be in the tens or even hundreds of thousands. The recording system must, therefore, have sufficient capacity to store up to tens of terabytes of data per day and the cable system must also be able to transfer the data rates for that survey. Although some cableless systems do not have concerns with data transmission thresholds, any breaks or drop of signal connection can be problematic.

One last point to be made about the onshore acquisition is the huge logistical undertaking of sustaining not only the seismic equipment but the vehicles and acquisition crew. Many land surveys are carried out in remote and hostile locations (i.e., deserts or mountainous regions) which require the setup (and removal) of a fit-for-purpose base camp. Huge quantities of water and fuel supplies as well as secure and weatherproof storage for all the equipment are necessary. Resupplying can be particularly difficult for specialist equipment or vehicles parts, so well-equipped workshops are also essential. Finally, power supply is essential and is a key service that must be preserved throughout the duration of a survey campaign.

3.7.2   Source Shooting

In areas inaccessible to Vibroseis vehicles, such as mountainous terrain, thick unconsolidated soils or areas with very dense vegetation, placing explosives in shot holes typically 5–25 m depth is a viable alternative. Sometimes a survey uses a combination of both Vibroseis and explosive sources, though explosives have several disadvantages namely high costs, safety concerns and that it is far more labour intensive.

Vibrator trucks can traverse a wide range of locations and terrain to deliver acoustic energy to a survey area. Older surveys may have employed a single vibrator but, like the modern marine surveys using multiple source arrays, it is common to use anywhere from four to twelve trucks simultaneously when shooting a large three-dimensional onshore survey. Trucks can be separated into individual groups referred to as fleets. Each fleet can be deployed to cover a different area. The vibrators have full GPS systems allowing the exact position of each shot point to be recorded as well as the movement of the truck.

Once receivers have been placed in the survey area the fleets travel to pre-specified source point coordinates where they stop and perform a sweep which may last between eight to twenty seconds depending on the survey design parameters. Older conventional surveys typically used two vibrator trucks, though each truck could only move to the next source point when the other truck had finished its sweep and the signal was recorded. However, modern acquisition techniques such as slip sweeps allow all the trucks in a vibrator fleet to move continuously from shot point to shot point, which hugely increases productivity and number of shots made per hour. Vibrator trucks are kept a fixed distances from each other and follow a pre-determined acquisition pattern time of shooting and moving. This method does create overlapping shot records and produces additional noise from the moving trucks, but modern seismic data processing is able to handle these issues. The time it takes for a truck to get to the next shot location is called the slip time.

In the past, trucks had to actively communicate with each other for position and status (i.e., the times of shooting and transit to the next shot position). Often communication abilities could be hampered by radio shadow zones caused by the terrain. With modern digital communication and positioning systems, it is possible to monitor and record positions and timings of all source trucks in real time and have the recording system continuously on ready to record the source shots.

With wireless technology vibrator trucks can easily be synchronized in a method known as slip sweep acquisition. Multiple source trucks are used in proximity to each other all running the same sweep pattern to amplify the signal. Rather than synchronized seismic source sweeps, groups of vibrator trucks run small, staggered sweep cycles as they move from one source location to another. Slip sweeps can be characterized by the same starting frequency, end frequency and rate of frequency change. This can be timed so that continuous vibrating is achieved without the need to waiting for a complete shot record to be finished.

Another acquisition method known as a simultaneous sweep is employed when trucks simply shoot the survey independent of the other trucks in the fleet and do not follow a pre-determined slip time. This greatly increases the operation efficiency. Distance-separated simultaneous sweeping is another variation that involves broadly spaced shot positions that all the trucks in a fleet move and shoot together. This can create tremendous energy and cover wide survey areas in a small amount of time. Figure 3.25 shows a comparison between the source deployment methods (Mougenot and Boucard 2012; Crook 2018).

3.8   Seismic Survey Suitability

When it comes to seismic surveys, either offshore or onshore, the most common question a geoscientist will face is: "Is this survey data optimum for the area and depth of interest?". Often the seismic data will have been acquired several years prior and in some circumstances can be much older and originally acquired without the current interval of interest in mind. The following options can be explored when faced with this question:

(a) Do nothing—The data are modern and entirely suitable for the project objectives. Or the budget of the project does not allow

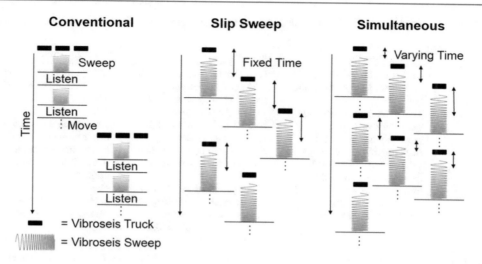

Fig. 3.25 Methods for Vibroseis source operation geometries (Crook 2018)

anything else to be done with the data and, so, maximum value needs to be squeezed out of what is available with heavy assumptions and caveats applied to geological and geophysical interpretation;

(b) Re-processing—The data are most likely old or poorly processed and a case can be made to target specific issues in the data that could be easily improved by a re-processing project to meet the project needs and bring additional value;

(c) Data purchase—The data are most likely old or poorly processed and it is questionable if the cost of re-processing can sufficiently improve the data to meet the project requirements, but the value and budget for a new acquisition are not sufficient. The option may exist to purchase multi-client (non-proprietary) data from a vendor which has been previously shot over the area of interest;

(d) New acquisition—There is no current data available over the area (frontier acreage) or the data are entirely unsuitable in quality. Reprocessing will not bring the required image improvement to show value, or the current data does not fully cover the area of interest. For a large discovery or development, the current available seismic data are insufficient to optimize development drilling

and production activities. This proprietary option comes with additional benefit over multi-client data listed in Table 3.2.

All operator and production companies will carry out some level of seismic data health checks to truly understand if a data set is fit for purpose before embarking on any subsurface projects. This practice is highly recommended as it manages expectation of the level of accuracy and detail of any geological interpretations made using the data.

A common example of how survey coverage over an exploration area could appear is shown in Fig. 3.26. As with most mature areas, two-dimensional seismic data are usually available and will be the oldest data. Additional three-dimensional surveys are shot in an attempt to tie into existing seismic data that has hinted at something of interest. The technology and complexity of the survey geometries increase as the survey technology becomes more modern. Prospects spanning multiple survey vintages can be inconsistent in their seismic response and so care during processing and interpretation is required. It is common for some parts of a prospect to look more favourable on one survey than another and a tendency to dismiss the more broken up or irregular response is all too common. Often the

Table 3.2 Comparison of the advantages and disadvantages of data re-processing, multi-client and proprietary seismic data

	Data re-processing	Multi-client survey	Proprietary survey
Cost	• Cost is generally cheaper than the other options in the table • The number of processing workflows will affect cost especially advanced pre-stack migration and complex velocity model build (FWI) • Getting access to field tapes from a survey contractor or processing company acquisition can be expensive	• Cost can vary depending on whether the latest/most advanced product is chosen • Field tapes and gathers can be a lot more expensive than stacked data volumes • Still likely to be 10–20 times cheaper than a proprietary survey • If not yet acquired (so only underwritten), the timing of the survey is up to the contractor and not the data purchasing company	• Always more expensive than re-processing or buying off-the-shelf multi-client survey data • Unforeseen costs can be a factor as survey acquisition is not straightforward • A lot of work is required upfront of the survey in the planning stages (i.e., permitting, EIA report, HSE, illumination studies).
Technology	• Similar to the multi-client survey product, although the source signature will remain as that in the original survey • Some workflows such as de-ghosting will require specific data (i.e., navigation and sea state). If these data are unavailable the solution can be sub-optimal compared to the multi-client option	• You usually have a choice, especially with large vendors, to buy from a range of older technology surveys and newer surveys • Products are unlikely to have the most cutting-edge acquisition and processing technology applied • It is very common that the data will be improved by bespoke modern re-processing post-purchase	• You can choose the absolute best the industry has to offer in both acquisition and processing technology • The ability to integrate this into the design phase and customize to suit your aims is probably the biggest strength of proprietary surveys • Best technology product that competitors are unlikely to have access to
Speed	• Slower than receiving multi-client data but quicker than new acquisition • Projects can drag if decisions are not made on intermediate products or the client wants extensive testing or to re-visit previously completed workflow steps	• Very quick, just requires data to be copied on hard disc or uploaded to FTP site • Some items such as pre-stack gathers, raw field tapes might take much longer to transcribe onto hard disc and at additional cost	• Acquisitions are planned to a certain time frame but factors such as weather, vessel issues and, equipment issues can add anywhere from +10 to +50% extra • Processing a new survey is not a linear workflow and rarely do final products get delivered within the planned timeframe
Quality of solution	• Can produce a much better result for a specified target than off-the-shelf multi-client data • Vintage data can receive an incredible uplift in data quality with the application of modern reprocessing workflows	• Most multi-client surveys are not acquired or processed with a focus on a particular subsurface interval and, therefore, are a general solution • 90% of the time additional post-migration conditioning to gathers or stacks for your target interval will need to be carried out	• Should provide the best quality solution possible • QC of vessel operations, acquisition tolerances and raw data is critical to achieving the best product • A highly experienced processing team is a key component needed to reach the best solution
Exclusivity	• Old data needing re-processing is likely to be available to competitors	• Data are freely available to potential competitors who may be looking at the same area	• Usually, the data will stay proprietary for a number of years (region dependent) giving the data owner

(continued)

Table 3.2 (continued)

	Data re-processing	Multi-client survey	Proprietary survey
	• High-quality re-processing can uncover important, unseen features and details for an area		insights and advantages over competitors

Fig. 3.26 An example of the variety of seismic surveys from a single area

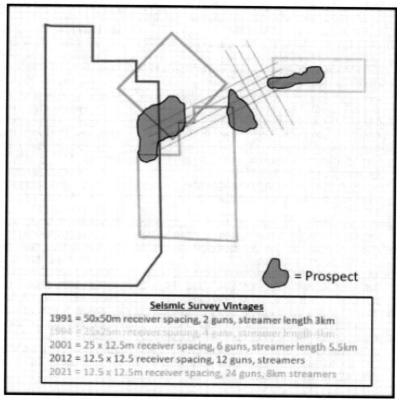

= Prospect

Seismic Survey Vintages
1991 = 50x50m receiver spacing, 2 guns, streamer length 3km
1994 = 25x25m receiver spacing, 4 guns, streamer length 4km
2001 = 25 x 12.5m receiver spacing, 6 guns, streamer length 5.5km
2012 = 12.5 x 12.5 receiver spacing, 12 guns, streamers
2021 = 12.5 x 12.5m receiver spacing, 24 guns, 8km streamers

better response is seen in the more modern of the two surveys falsely re-enforcing the idea that "newer is better". However, this can be a dangerous assumption.

The use of two-dimensional seismic survey geometries within the energy industry is becoming less and less common. New two-dimensional acquisitions are most likely used in areas that are underexplored and, as such, the investment in a more expensive survey cannot be justified. Another reason to shoot a new two-dimensional survey is to cover a large area to get a regional understanding in a short time frame. Economics clearly underpin this choice as the uplift of shooting a three-dimensional survey can be substantial.

References

Abma R, Ross A (2013) Popcorn shooting: sparse inversion and the distribution of airgun array energy over time. In: 83th annual international meeting, SEG, expanded abstracts, pp. 31–35. https://doi.org/10.1190/segam2013-0592.1

Brown AR (2011) Interpretation of three-dimensional seismic data. Soc Explor Geophys Am Assoc Petrol Geol

Chaouch A, Mari J (2006) 3-D land seismic surveys: definition of geophysical parameters. Oil & Gas Sci Technol Rev IFP 61(5):611–630. https://doi.org/10.2516/ogst:2006002

Chmiel M, Mordret A, Boué P, Brenguier F, Lecocq T, Courbis R, Hollis D, Campman X, Romijn R, Van der Veen W (2019) Ambient noise multimode Rayleigh and love wave tomography to determine the shear velocity structure above the Groningen gas field. Geophys J Int 218(3):1781–1795

Crook A (2018) Seismic acquisition innovations applied in Canada. CSEG Recorder 43(05)

Curtis A, Gerstoft P, Sato H, Snieder R, Wapenaar K (2006) Seismic interferometry—turning noise into signal. Lead Edge 25(9):1082–1092. https://doi.org/10.1190/1.2349814

Doll W (1984) Kinetics of crack tip craze zone before and during fracture. Polym Eng Sci 24(10):798–808

Edelmann HAK (1981) SHOVER * shear-wave generation by vibration orthogonal to the polarization **. Geophys Prospect 29(4):1365–2478

Ewing M, Vine A (1938) Deep-sea measurements without wires and cables. Trans, Am Geophy Union, Part 1:248–251

Grazier V, (2009) The response to complex ground motions of seismometers with galperin sensor configuration: bulletin of the Seismological Society of America. 99, no. 2B, 1366–1377

Hornby BE, Yu J (2007) Interferometric imaging of a salt flank using walkaway VSP data. Lead Edge 26 (6):760–763. https://doi.org/10.1190/1.2748493

Hovland V (2016) Transforming ocean bottom seismic technology into an exploration tool. First Break 34(11)

Kapotas S, Tselentis G-A, Martakis N (2003) Case study in NW Greece of passive seismic tomography: a new tool for hydrocarbon exploration. First Break 21 (1007):37–42

LandrØ M, Amundsen L (2010) Marine Seismic Sources Part 1. GEO ExPro Vol. 7, No.1

Landrø M, Amundsen L, Langhammer J (2013) Repeatability issues of high-frequency signals emitted by airgun arrays. Geophysics 78(6):19P https://doi.org/10.1190/geo2013-0142.1

Li Q, Slopey S, Rollins F, Billette F, Udengaard C, Thompson BJ (2019) Leading a new deep water OBN acquisition era: two 2017–2018 GoM OBN surveys. In: SEG international exposition and 89th annual meeting. https://doi.org/10.1190/segam2019-3216437.1

MacLeod MK, Hanson RA, Bell CR, McHugo S (1999) The Alba field ocean bottom cable seismic survey: impact on development. Lead Edge 18:1306–1312

Meunier J (2011) Seismic acquisition from yesterday to tomorrow. Soc Explor Geophys, Tulsa, OK

Mougenot D, Boucard D (2012) How technology drives high-productivity vibroseis: a historical perspective. Geophys Prospect 60:602–607

Mougenot D (2004) How digital sensors compare to geophones? SEG Techn Program Expand Abstr 5–8

Prabowo BS, Ry RV, Nugraha AD, Siska K (2017) Hydrocarbon prospect derived from attributes analysis on low-frequency passive seismic survey: a case study from Kalimantan, Indonesia. IOP Conf Ser Earth Environ Sci 62

Stroujkova A, Leidig M, Bonner JL (2015) Effect of the detonation velocity of explosives on seismic radiation. Bull Seismol Soc Am 105(2A):599–611. https://doi.org/10.1785/0120140115

Wallace J (2018) Multiple source acquisition for use in 4D marine seismic. First Break 36(11):77–82

Zachariadis RG, Thomason HB, Teague HE (1983) Ocean bottom seismometers in seismic exploration surveys: planning and operations. 53rd Ann Meet SEG Expand Abstr S15.6, 468–470

Ziolkowski A (1984) The delft airgun experiment: first break 2(6):9–18

Processing Essentials

4

Seismic processing is a specialist discipline that sits within the broader label of geophysics. The simple aim of seismic processing is to take the raw recorded seismic data and to turn it into an interpretable image to show details about the subsurface geology. Although seismic interpreters do not generally need to process seismic data, it is very likely they may be involved in a processing project. Knowing about the advantages and pitfalls of certain processing steps can be useful to distinguish between real seismic events and artefacts. This chapter focuses on specific processing workflows that are likely to have the largest impact on subsequent steps of the geo-modelling workflow and what the interpreter should look out for.

For all seismic acquisition projects, a series of QC checks are performed in the field to ensure that the parameters set in the survey specifications are met, otherwise a reshoot of a particular line, or section, may be considered. Marine surveys (Sect. 3.4), for example, carry out extensive QCs of navigation data such as the position of the vessel(s) during shots, gun depth and spatial movement of the streamer during acquisition. This is all done via GPS and has sensitivities on some systems of less than one meter. Source QC looks at the consistency of pressure in the air guns as well as the signature strength and shape of the pulse being produced. Over the decades onboard processing has become increasingly advanced, with some companies offering fast-track processing flows producing fully migrated

(time or depth) de-ghosted broadband products, which upon completion of the survey can be sent onshore via satellite link. Most onboard survey QCs are applicable to both streamer and ocean bottom acquisition, but not all. Limited field processing is carried out for OBN or OBC with everything done onshore.

Although onboard processing products are becoming more sophisticated and give greater insight to the quality of data and specific areas of interest in the subsurface, it is onshore processing that will ultimately be used to produce the final image used by interpreters and geoscientists. Due to the level of complexity in a modern processing workflow, the seismic interpreter and processor should work together to ensure that the workflow applied to the data is fit for purpose based on the aim(s) of the survey acquisition or reprocessing objectives.

The aim of this chapter is to highlight some key processing topics and workflows that impact the fidelity and interpretability of the seismic data, which are of particular importance to the interpreter. Our focus, for the most part, is on marine seismic processing workflows and processes. Key workflows related to onshore processing will not be reviewed, however, clear synergies between some processing steps provides the reader guidance for both marine and land processing. For detailed theory and methodology resources for seismic processing the reader can refer to Yilmaz (2001), Sheriff (1991) and Hill and Rüger (2020).

© The Author(s), under exclusive license to Springer Nature Switzerland AG 2022
T. Tylor-Jones and L. Azevedo, *A Practical Guide to Seismic Reservoir Characterization*,
Advances in Oil and Gas Exploration & Production, https://doi.org/10.1007/978-3-030-99854-7_4

Vintage (from the 1980s) streamer processing workflow	Modern OBN processing workflow
1. Transcription	1. Source/receiver re-positioning
2. S.O.D. Correction	2. Resample and low-cut filtering
3. Signature deconvolutions	3. Q-compensation
4. Resample	4. Denoise
5. Multi-channel filtering	5. Statics
6. CMP gather	6. Shot regularization
7. De-multiple	7. De-signature
8. Dip moveout	8. PZ calibration and summation
9. NMO correction	9. De-multiple
10. Mute	10. Surface consistent corrections
11. Amp equalisation	11. 4D binning and 3D regularisation
12. Deconvolution	12. Pre-stack depth migration
13. Migration	13. Pre-stack post-migration processing (NMO)
14. Spectral shaping	14. Post-stack post-migration processing
15. Bandpass filtering	15. Matching
16. Equalization	16. Stacking
17. Stacking	

Fig. 4.1 Comparison between vintage and modern marine offshore processing flows

Figure 4.1 shows the comparison of a generic marine processing flow from the 1980s with that of a typical modern marine OBN processing project. There are some common processing steps at the start of each flow (e.g., data formatting and de-signature), further into the flow (e.g., de-multiple and NMO) and at the end (i.e., stacking), and the aims of the processing steps are similar between the two flows, however the methods and algorithms applied are vastly different. Even within modern processing flows the implementation of the same processing steps often vary between different processing companies.

4.1 Wavelet Processing: De-Bubble, De-Ghost, Zero Phasing

Source de-signature in marine processing looks at the seismic pulse produced from the source that has interacted with the water column and subsurface to produce the field record recorded by the receivers. In the case of marine offshore acquisition, the source is produced by airgun arrays (Fig. 3.2) and its pulse is embedded in the seismic data and can be isolated as discrete wavelets (Chap. 3). Although the specific characteristics of the source wavelet signature produced by the airguns are known upon firing the airguns, subsequent physical interactions change the signature's properties.

The recorded signature of an airgun has three main components: the direct arrival from the main pressure release when the gun port is opened; the source ghost from the energy which travels upward when the gun is fired and interacts with the water–air boundary; and the bubble pulse produced by the expansion and collapse of the air bubble (Dragoset 2000). Figure 4.2 shows a synthetic version of what these components might look like for the pressure signature of a single airgun recorded from a near-field hydrophone (i.e., a hydrophone deployed few meters below the airgun). Source de-signature is used to address three main issues: removal of the bubble pulses (Fig. 3.4), removal of the source ghost (Fig. 3.6) and source wavelet zero phasing. The characteristics of the direct arrival pulse should be a high energy amplitude and a short time duration which makes its appearance narrow.

Fig. 4.2 Signature of a single airgun recorded by a hydrophone 300 m beneath the source

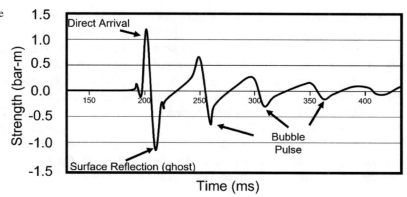

4.1.1 Far-Field Signature

The far-field pressure signature measurement can be used to calculate estimates of the levels of acoustic energy that arrive at the hydrophones at different offsets. To accurately capture the combined energy generated by all the guns in a single array, represented in a single shot, a hydrophone is used to make a recording of the source pulse from the gun array. In general, the hydrophone must make the recording at a depth of approximately 200–300 m below the gun array, to account for any constructive interactions of the various bubbles. This measurement position is often referred to as the far-field point. An important source energy QC known as the peak-to-bubble ratio (PBR) can be calculated from this measurement and should be as high as possible to indicate that the direct arrival is close to an ideal pulse. If the interpreter knows that the PBR in the data set is lower than expected, issues with wave primary reflection amplitude strength during the processing and imaging could lead to a challenging interpretation of the data set.

Once the far-field pressure signature measurement is obtained it can be used to generate an understanding of the pressure energy distribution within the survey by calculating a theoretical point source ∼1 m vertically beneath the source array depth. This is called the nominal point source (Ziolkowski et al. 1983). For example, if the nominal point source has a pressure signature of 240 bar-m and the nearest hydrophone is 80 m then we can expect 3 bar (240/80 = 3) of

pressure at that receiver. The theoretical point-source energy estimation has uncertainty associated with it and, as such, the energy calculated to reach individual hydrophones at distances from the source is widely accepted to be an over-estimation.

Another important QC during de-signature is the peak-to-peak measurement, which is the difference between the strength of the first arrival pressure energy, calculated using the nominal point-source level, and ghost arrival energy recorded in the far-field pressure signature. Understanding the difference in amplitude strength between the two pressures is important for understanding the source output of the survey (Landrø and Amundsen 2010).

4.1.2 De-Bubble

The airgun array is designed to create a short, sharp, high-energy source signature by combining a variety of airgun bubbles to give the best possible PBR value. However, there is often residual bubble energy in the signature that negatively impacts the final seismic image and, therefore, requires removal (Fig. 4.2).

The source wavelet embedded within the recorded seismic data is the product of the convolution of the far-field source signature with the earth's filtering effects. The oscillations effect the source wavelet in the water column before it travels through the subsurface and is recorded at the receiver. These oscillations can be observed on seismic shot records as additional reflection events which can mask and interfere with the recorded direct arrivals. Leaving residual bubble energy in the seismic data results in unwanted reverberation energy in the final seismic image. From an interpretation perspective an effective de-bubble process is important because it removes reverberations which can affect the clarity and resolution of a seismic image. If the processor can correctly estimate the bubble effect and remove it, the data will be free from its effect.

Early de-bubble techniques took a deterministic approach and used matching filters to correct bubble reverberation effects on the source wavelet. These techniques used cross-correlation of the known gun signature, which came from either prior knowledge or wavelet estimation with the recorded data. These methods shaped the embedded source wavelet so the data would contain the source signature's autocorrelation function as a basic wavelet. This process had additional benefits of increasing signal-to-noise and making the data zero-phase (Wood et al. 1978). Additionally, the wavelet would often be narrowed using a filter which improved the overall resolution of the data. These early techniques struggled to produce a proper estimate of the far-field signature.

Modern de-bubble techniques focus on getting the best estimate of the far-field signature. We have explained how a near-field hydrophone (NFH) can be used to help estimate the far-field signature and nominal point source, however issues with the far-field signature estimation can still arise even with NFH measurements. Sometimes guns will not fire (drop-out) and the pressure can vary from shot-to-shot. Additionally, external factors such as physical condition of the sea and weather may lead to differences in a modelled far-field signature.

There are various techniques for estimating far-field signature from data sets without NFH data, for example using the sea-bottom reflectors to estimate the source signature which is in turn used to estimate an estimated source (Scholtz et al. 2015), as well as extracting a far-field signature from the direct arrivals (Davion and Poole 2015). Additionally, the far-field signature can be synthetically modelled using information about the gun array configuration (Landrø 1992). These processes can then model the far-field signature to produce an operator for the de-bubble process (Casasanta et al. 2020). With the recent advances in machine learning there are ever more and evolving variations on new methods such as deep learning and neural networks for far-field source signature (de Jonge et al. 2022).

4.1.3 Source De-Ghost and Zero-Phase

The source ghost is the caused by energy that travels directly up from the airgun and is reflected downwards from the sea surface (Sect. 3.4 and Chap. 6). The ghost is recorded at almost the exact same time as the source pulse, but with the opposite polarity. The source ghost can interfere with the ability to record the full range of frequencies in the seismic data and should, therefore, be removed. The de-signature measurement from the NFH captures the source ghost which is used in the PBR QC analysis. Source ghost removal can be carried out in the de-signature step of processing flow, though bear in mind there is also a receiver ghost which is not removed using this process and needs to be addressed in the de-ghosting workflow that will be applied to the data later in the processing flow.

Most seismic data (both marine and land) are changed during processing to become zero-phase, which results in maximum acoustic amplitude being focussed on the centre of peaks and trough. The action of making the data zero-phase allows the interpreter to understand what the peaks and troughs they pick represent in the data. (i.e., a reflection boundary). Picking horizons on seismic data not in zero-phase means that the event picked in the amplitude minimum or maximum is not the true interface and, consequently, any amplitude extracted on the event might not be representative of the impedance contrast at that boundary.

The phase of the seismic data is adjusted during the de-signature stage of data processing. The interpreter should not assume the seismic data are zero-phase unless they have been involved in the processing and seen evidence of zero phasing. However, there is a quick way interpreter can easily identify if the data are zero-phase.

The wavelet estimation performed during well tying (Chap. 6) should be able to produce a zero-phase wavelet, or one very close to it. Consistent large phase rotations (>10°–15°) in wavelets taken at numerous wells in areas of seismic data with high signal-to-noise may indicate that either the zero-phasing was not applied successfully or another process has affected the phase of the data post-de-signature. Difficulty in aligning multiple strong seismic events, within a reasonable time window, during the well tie is also a sign of poor zero-phasing and the interpreter should refrain from using time-depth stretch-and-squeeze to force events to match in the well tie. Should the interpreter identify that the zero-phasing applied is not sufficient an operator with the correct phase can be designed and applied to a final migrated volume. As this is a relatively simple solution to the problem it becomes more important that the interpreter identifies the phase issue in the first place.

4.2   Noise Removal

Noise represents signal in the data that do not give information about the subsurface and actually mask the useful reflections events. Therefore, effective removal of noise should not damage the primary energy or lead to loss of information about the subsurface. Although noise suppression is a major consideration when designing seismic acquisition, some noise is always expected to be present in the recorded data. The mechanisms and sensitivities of both hydrophones and geophones are described in Chap. 3 along with the concept of noise.

The common way to classify noise is to relate it, if possible, to the source from which it is generated. Elboth et al. (2008) separate noise into background noise which includes natural (e.g., swell, wind), man-made sources (e.g., platform and vessels activities), source-generated noise connected to the release of seismic energy and its ray paths (e.g., multiples, scattered waves and direct waves) and instrument and survey equipment noise, which is associated with all of the survey equipment in the field. In seismic data, noise comes in two main types: coherent and incoherent. This distinction results in differing processing methods for removal and attenuation.

For the processor, denoise is often a trade-off between removing too much noise, giving the data an almost synthetic look, versus leaving in too much noise but capturing potentially important subtle inflections. What appears as noise may actually be a legitimate lithology response, such as the end of a sedimentary system or coarse breccia sediment load. However, these responses would tend to be localised in the data.

Applying denoise techniques early in the processing flow (pre-migration) is advantageous and will improve the migration results. Many standard time-migration algorithms assume all the data are primary and will try to migrate the noise. Post-migration noise removal is all about making reflection events cleaner to allow for more accurate interpretation.

In any type of noise removal process the general objective is to transform the data into a domain where it can be separated from the primary signal. The noise can then be removed before transforming the data back to its original domain. Another approach is to use observations during the domain transform to identify characteristics around things such as frequency, dip, moveout or amplitude, which can be used to create a filter or other tool to remove noise from the primary signal in its original domain. Noise removal is an extensive area of research, and the following sections are not exhaustive; a more comprehensive guide to key noise types is provided in Hlebnikov et al. (2021).

4.2.1   Data Sampling

The sampling of the data needs to be sufficiently dense, in both space and time domains, to produce coherent reflection events for final imaging (i.e., migration). Issues around variable data density are often experienced at the edges of a survey area, where the data fold may be lower than the main part of the survey. At the survey edges poorly defined reflectors and migration 'smiles' are often observed, which are indicative of the poorly migrated reflections. The interpreter should be aware of the migration aperture in relation to the survey extents and spatial coverage for the full-fold area, which will be correctly migrated. Coarsely sampled data may also cause problems with correctly migrated high angle dips, and the migration may not produce a reflector that is clear or interpretable.

In Chap. 2 we discussed the concept of aliasing and under sampling the signal. Under sampling can also occur when the acquisition survey is designed with sparse line spacing or shot timing, which results in poorly imaged primary events with weak amplitudes that are easily distorted by shallow diffraction multiples and noise. Furthermore, under sampling affects multiples, as well as primaries, which if under sampled can appear as noise that is difficult to remove.

Poor illumination of the subsurface, for example below salt features and around large dipping faults, can also lead to noisy data because the ratio of seismic signal is greatly reduced compared to the inherent noise in the recorded data. Shallow channels that scour the seabed can in some circumstances be observed to cut into the subsurface several hundred meters. These channels can have the effect of attenuating the seismic signal directly below them over long depths/TWT intervals. The channels can also act as uneven multiple generators adding further noise into the data. Therefore, it is always good practice to spend time mapping a seabed horizon and observe the over-burden in the seismic data within the area of interest.

4.2.2   Filters, Transforms and Stacking

When it comes to noise attenuation a huge variety of methods and techniques is available at the processing workstation. However, there are three main techniques used in almost every processing sequence.

Filters

A simple way to improve seismic data is filtering to remove unwanted noise. Filters work by removing data that falls within the user-defined range for a particular property. Spectral filtering is widely used to remove unwanted noise based on the frequency band it occupies. Another common method is F-k filtering, which operates by removing specific frequencies in both time and space domains. However, this type of filter becomes less effective when both the noise and signal occupy similar frequency ranges.

Some filters operate using spatial variation to identify and remove unwanted parts of the data. The F-X (frequency-space) deconvolution filter (Gulunay 1986) is applied for frequency and distance and is referred to as a predictive filter. The filter is applied in different directions across seismic traces and tries to predict the signal in the neighbour trace in the frequency domain. The method was first proposed by Canales (1984) using a method based on spatial signal prediction. The F-X method was added to include and combine time–space (T-X) for attenuating random noise by linear predicting filters (Abma and Claerbout 1995).

A more specialized filtering technique used for targeting incoherent noise is low-rank or rank-reduction denoise. This is a varied group of filters that use matrix factorization and tensor decomposition. These methods can be particularly good for the attenuation of random noise that is Gaussian in nature (Chen and Sacchi 2013). The technique is applied in the frequency domain but also uses multiple spatial domains. The Cadzow filter is a commonly used rank-reduction denoise filter (Tickett 2008).

The downside to filtering is the potential to damage the primary signal at the same time as attenuating the noise. Applying filtering techniques to the data after migration is commonly used as a method to remove incoherent noise. Filters such as median, or structurally orientated dip filters, can be particularly effective at removing noise after migration. Testing filter parameters like window size/trace radius, filter strength and number of iterations of filtering applied should be carried out by the interpreter to avoid over-filtering the data.

Transforms

Another widely used method to deal with noise in seismic data is to transform the data into different data domains (i.e., frequency and offset) to maximize the separation of the noise and signal. The filters mentioned above technically use transforms prior to filtering but most of the signal and noise separation is achieved by the application of the filter, not by the separation of the transform. When using a transform we try to select the noise in the transformed domain, bring it back to the original domain and subtract the noise there. This avoids any unwanted loss of primary data due to the forward and inverse transforms.

Sparsity noise reduction methods involve the data being transformed into a sparsity domain where the signal and noise can be separated. The method relies on the amplitude and location difference between the signal and noise within the transform domain. Wavelet transforms are commonly used for sparsity noise reduction because of their ability to represent non-stationary signal. There are different types of wavelets transform used for this purpose (e.g., bandlets, contourlets, curvelets, directionlets, shearlets, seislet) (Fomel 2006), with the commonality of exploiting directional characteristics of the signal within the data. Curvelet based transforms have proved effective as a high sparsity transform with little overlap between signal and noise, therefore causing minimal damage to the primary signal. Curvelets can also be used to tackle coherent noise. The seislet is considered a very high sparsity transform and is a wavelet transform specifically tailored to seismic data; it follows patterns of the seismic events based on slope and frequency (Liu and Fomel 2010).

Stacking

Stacking effectively deals with noise in both marine and onshore seismic data. Stacking is commonly done pre-migration on normal moveout (NMO) corrected CMP gathers or for post-migration stacking is carried out on migrated gathers, depending on the type of migration approach being used. By simply summing multiple seismic traces together from the same CMP location and across offsets angles, primary events become overlaid due to their similar spatial positions trace by trace. This has the effect of strengthening the primary reflection amplitudes. The noise, which tends to be random and unaligned from trace-to-trace, does not align or overlay and therefore the amplifying effect of the stack is negligible. As such, the enhancing effect of the stack increases the signal-to-noise ratio. There are many variations proposed for stacking that make use of both spatial and temporal information to effectively attenuate incoherent and random noise.

Common image point gathers can help with velocity analysis after migration by validating the correct positioning of dipping reflectors and avoid issues such as poorly focused reflectors and mis-positioned events. To create this type of gather a velocity model is required to be applied along with the shot-receiver geometry. Several image point gathers can be migrated with different velocities to help refine the correct stacking velocities required or to identify residual velocity correction. Image point gathers are synonymous with CDP. The two common processes associated with stacking velocities and residual velocity correction are NMO and RMO.

NMO utilizes a constructed velocity model to correct the moveout effect (δt). The moveout effect is defined as the difference in the arrival time of the different offsets of traces within the same CDP gather and having a greater than zero offset (Fig. 4.3). Seismic gathers that are not corrected will show hyperbolic seismic events

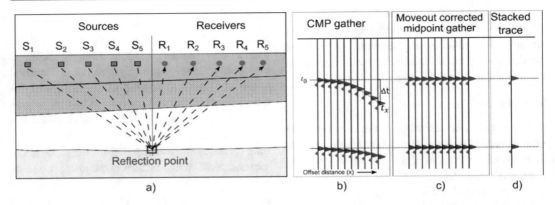

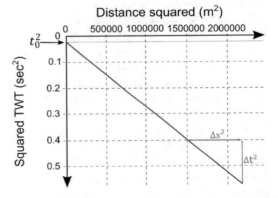

Fig. 4.3 **a** CMP reflection location; **b** CMP gather without NMO correction; **c** NMO corrected CMP gather; **d** stacked seismic trace

with increasing source-receiver offset. This will negatively affect the stack (i.e., summing horizontally all traces of a given CDP gather).

The reference position for each event in the CDP gather is the TWT of that seismic event at zero offset (t_0). If the NMO is applied correctly then all the reflection events will have the same TWT no matter what the source-receiver distance. This correction assumes that the seismic traces at a given CDP gather have been recorded for zero source-receiver offsets so they are flat and can be eventually stacked. The NMO correction tries iteratively to estimate the velocities' profiles that distort the different offset traces' arrival times and remove this distortion to match the zero-offset trace which has the least travel time distortion. Figure 4.3b shows a diagram with source-offset distance x, the vertical TWT at zero offset t_0 and a parabolic best-fit line going through the events the TWT angled offset ray path t_x. What we want to do is shift all events with a given time (Δt) with source-receiver offset greater than zero. This is represented by:

$$\Delta t = t_x - t_0. \tag{4.1}$$

So, how do we come up with the NMO velocities based on the known values for T_x–T_0? Fig. 4.3b shows two reflection events where the shallow reflector has a higher angled hyperbola compared to the deeper reflector due to the slower velocities

Fig. 4.4 A t^2 versus x^2 domain

in the shallower one. These linear events are plotted as straight lines in a t^2 versus x^2 domain (Fig. 4.4). Using the linear relationship, now created between t^2 and x^2 of the reflection events allows measurements of slope which represents:

$$slope = \frac{1}{V_{nmo}^2}, \tag{4.2}$$

with V_{nmo}^2 simply being calculated:

$$V_{NMO}^2 = \frac{\Delta x^2}{\Delta t^2}, \tag{4.3}$$

by taking the square

$$V_{NMO} = \sqrt{\frac{\Delta x^2}{\Delta t^2}}. \qquad (4.4)$$

Therefore, the NMO velocity (V_{NMO}) required for the correction can be made for each layer overlying the reflection event. Once calculated for all the events and traces a NMO velocity model is generated. This is often referred to as a stacking velocity because the application of this velocity-based correction gets the data ready for stacking.

Residual moveout correction (RMO) can be applied to gathers to flatten events showing hyperbolic moveout with offset. There are a number of reasons for applying RMO. The velocities in the migration model were either too high causing events to dip downward or too low causing events to dip upward. This behaviour will be seen across all events in the gather. It is also possible to see events dipping up in the shallow and down in the deep or vice versa which may suggest an additional layer(s) needs to be defined in the migration velocities.

The easiest way to define whether RMO is required is a visual inspection of the gathers after NMO. It should be clear that they are not flat and exhibit residual shallow hyperbolic moveout. By observing an amplitude graph of a particular event in a gather, strong variation in amplitude with offset can be further evidence for applying RMO.

Both NMO and RMO depend on our ability to create a representative model of P-wave velocity in the subsurface that accurately represents the geology. For moveout correction, the main task of the processor is to generate a velocity model that captures these velocity boundaries with the right magnitude of velocity. Contributions from the seismic interpreter, geologist and petro physicist are all useful to produce the best outcome. As velocities are the critical point for the success of NMO correction, we review a few different types of velocities that we can derive from seismic data assuming horizontal reflection boundaries and constant velocity values being assigned to coarse geological interval layers (Fig. 4.5).

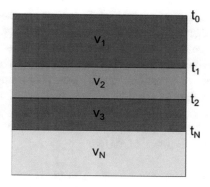

Fig. 4.5 Schematic representation of N geological layers with constant velocity layers

When dealing with multi-azimuth (MAZ) surveys, data stacking can become challenging as there are several different data sets in use. However, stacking in a MAZ survey is effective at dealing with both noise, due to increased fold, and suppression of multiple energy and diffractions. Several different stacking methods can be applied the simplest of which involves stacking individual azimuth sector migrated volumes together. Another approach is to create what is referred to as a super gather which involves collecting traces based on similar offsets and/or angles from neighbouring CMPs. The super gathers are then stacked together to improve signal-to-noise. This technique can also be used in non-MAZ surveys. A more advanced method, like a weighted stack, assigns appropriate weights to each azimuth sector gather or stack post-migration which gives priority to certain azimuths during the stacking. This may be done because features such as faults or geological boundaries are better imaged in particular azimuths more than others. This type of weighting should be done after careful QC of data for each individual azimuth.

Although stacking is a robust tool, a sense check should be made of the data's usable offset range that went into making a final post-migration stack. This ensures that no data has been included that degrades the signal of the stack (i.e., noisy near offset/angle traces, unflattened data at long offsets/angles), which is particularly important when generating multiple partial stacks. In this case, the geoscientist should only settle on final partial-stack ranges after reviewing several different combinations using visual observations from the post-migration gather data as a guide.

4.2.3 Coherent Noise

Coherent noise in seismic data is visually easier to identify than incoherent noise because it aligns trace-to-trace and appears as a single discrete and measurable event(s) that can be visually traced on recorded data. If coherent noise is not properly sampled it might appear as random noise. Coherent noise is a particular problem for the seismic interpreter because, unlike incoherent noise, the coherent noise events can appear as false primary signal reflection events, which may be misinterpreted as true reflections. Some coherent noise is very difficult to spatially differentiate from the primary signal, making it challenging to remove without risking damage to the primaries. For this reason, it is important that the seismic interpreter is aware of denoise QC stages of the processing and is comfortable with any trade-off between noise removal and damage to primary energy. There are several methods to remove coherent noise: separation through transforms; modelling the noise then subtracting from the data; and the use of inverse filters. Coherent noise can be sub-divided to include categories such as linear noise, ground roll, guided waves, diffractions, and reverberation.

4.2.4 Linear Noise

Linear noise is a visually descriptive term used for noise that appears as near straight lines, usually at high angle dips and high-frequencies, which crosscuts primary events. Linear noise is often present throughout a data set at the full range of TWTs and offsets, rather than being spatially isolated. The appearance in a seismic stacked section it appears as a noisy overlay on the primaries. High-frequency linear noise can have several possible origins. In marine acquisition, machinery or equipment along with cavitations from a vessel's propulsion may be the source. Loose brackets or vibrations from equipment on the streamer array are also possible candidates. For onshore surveys, repetitive high-frequency human activity or machinery are just a few potential sources. High angle linear noise is also commonly caused by multiple energy such as diffracted waves and free-surface multiples, which are discussed individually in this section.

Linear noise can also be low-frequency and vertical in nature, which can be just as problematic for the interpreter. Low-frequency linear noise has a longer wavelength and, thus, attenuates more slowly allowing it to penetrate deeper into the subsurface. Some low-frequency noise is very low-velocity (<1500 m/s) and may not penetrate the subsurface but be trapped near the sea surface. Low-frequency mechanical equipment is a good example of a source for this type of non-penetrating low-frequency noise. When low-frequency linear noise is present in seismic data, a stacked section will show long vertical striping throughout the data. Two commonly encountered types of low-frequency linear noise are swell noise and cable noise.

Movement of the sea surface during rough weather conditions causes swell noise in the recorded data by either inducing motion and deformation of the streamers or by generating seismic energy waves that interact with the seafloor. The effect of swell noise is particularly noticeable in shallow water surveys and manifests itself on shot records as long, broad, vertical (often) high amplitude streaks in the data. Due to its low-frequency, a simple low-cut or bandpass filter may be enough to remove the swell noise, however the preservation of usable low-frequencies in the data should always be a priority so the filter needs to be carefully designed.

Seismic streamer cables being pulled through the water also generate noise due to the tugging action from the vessel, which is referred to as strumming noise. Similar to that of swell noise, it is often either low-frequency or within a narrow frequency range and is mostly dealt with by the same low-cut or bandpass filtering applied to remove swell noise.

For coherent linear noise the removal techniques are varied and target different attributes. Transforming the data into other domains to separate the noise (e.g., F-k, Tau-P) are traditional solutions. The high-frequency nature of the noise and its amplitudes can also be used to help separate them from primary events. The difference in dip angle with offset is another method exploited for removal, however, because this type of noise crosscuts the primaries the removal has a high potential for damaging the primaries. Due to this risk the interpreter should visually review the removed noise to check for residual coherent primary signal. A simple but informative QC is to view the removed noise in isolation as a separate volume and to check for coherent primary events. Additionally, making a seismic stack during denoise testing is recommended to check for variations to reflections and any amplitudes of interest.

4.2.5   Guided Waves

Guided waves (Chap. 2) travel in a horizontal direction along shallow geological surfaces. The waves can contain information about shallow geology but have little information about the deeper subsurface, and so are often removed because their high amplitudes can obscure deeper reflections. These waves are very dispersive in nature which makes their effective attenuation challenging. However, where dispersive waves appear in the shallow parts of the data, they can be simply muted out from the data without damaging important information. F-k filtering domain, or mean filters, are common removal methods. More advanced methods focus on areas such as wave-theory based prediction to remove the part of the recorded signal which contains the surface waves (Ernst et al. 2002). Inversion based methods are also becoming more commonly used to more accurately remove guided waves (e.g., Douma et al. 2014; Stobbia et al. 2010).

4.2.6   Ground Roll

Ground roll is classified as source-generated noise caused by the vertical component of a particular type of Rayleigh wave travelling through shallow subsurface layers. Recorded ground roll noise is a common type of coherent noise in land seismic surveys, although marine surveys that use either OBN or OBC based receivers may also encounter ground roll. This type of noise is characterized as having a high amplitude, low-frequency and low-velocity and is dispersive in nature, which masks shallow reflections at short offset and deep reflections at longer offset (Torabi and Javaherian 2012) affecting many primary reflections within the data. Its high amplitudes mask weaker primary amplitudes that affect the ability to interpret the data. It is therefore a priority process for denoise prior to interpretation.

Ground roll can be mitigated during acquisition by either placing the source below the shallow layers (i.e., borehole explosives) or using receiver arrays to attenuate the surface waves and prioritise reflection energy. Multi-component receivers are also useful for trying to separate the ground roll using wave directionality. It is widely accepted that the majority of ground roll is dealt with during seismic processing. The removal techniques for ground roll are extensive and varied. Earlier methods involved applying filtering techniques to the seismic data, for example multichannel filters, deconvolution, frequency balancing, frequency filters and dip filtering. However, filters rely on the coherent noise being consistently sampled so where this is not the case residual noise can be left behind. Because ground roll is both low-frequency and dispersive filtering often fails to remove all the noise. Applying filters too aggressively to counteract this leads to the removal and damage of primary energy. Emerging approaches to remove ground roll favour methods like curvelet transform, which is a geometric multi-scale transformation with an optimal disperse representation of the seismic data (Souza and Loures 2009). Further reading on the application of curvelets for ground roll include Yarham et al. (2006) and

Naghizadeh and Sacchi (2018). The application of machine learning methods to attenuate ground roll is discussed in Yuan et al. (2020) and Pham and Li (2022).

4.2.7 Diffractions

Multiple diffractions can be included in coherent noise; however, a more comprehensive discussion is made later in this chapter because multiples do not purely fit the definition of what constitutes noise. Unlike the definition of noise, multiples do contain information about the subsurface and are useful in a range of applications to improve imaging.

Multiple diffractions can be a significant noise problem and are most prevalent in deep water surveys where complex near surface reflection sequences are present (Brittan and Wrench 2004). The water bottom irregularities act as a scattering surface which results in multiple diffractions that exhibit the following properties: strong amplitudes, non-periodic, complex moveout in CMP gathers spatially distributed in three-dimensions (Hargreaves and Wombell 2004). These multiples produce associated sidescatter energy, which has the appearance on a stacked section of linear high angle noise. Several methods can be attempted to remove this noise including filtering and transforms (e.g., F-k filtering, Tau-P transform, pre-stack dip-filtering and radon transform). Removal of multiple noise is particularly important as it can dominate seismic sections making subtle interpretation difficult. High angle noise can also interfere with seismic attribute extractions making them appear noisy and grainy in consistency.

4.2.8 Reverberations

Water-layer reverberations (i.e., free-surface multiples) are problematic in marine surveys. They are source-generated noise resulting from interaction between two interfaces, in this case the water surface and water bottom, although this can happen between any two interface layers. The water-layer reverberation occurs when a down going wave hits the water bottom and is reflected up toward the receivers near the sea surface. Some energy is recorded at the receivers as a primary reflection of the seabed, but some energy reaches the sea surface and is reflected down again. This energy is reflected multiple times within the water column before finally being recorded.

Water-layer reverberations can mask both shallow and deep primary reflections within seismic data. The noise can appear in different ways in a CMP gather such as long trains of short-pulse high amplitude energy, horizontal crosscutting energy resembling the sea floor or larger clouds of noise. When surveys are shot in areas where the seafloor has a large acoustic contrast with the water column, more energy is reflected upward leading to the generation of stronger reverberations. Figure 4.6a shows the ray paths that generate the water-layer reverberation.

Shallow peg-leg multiples can also be included in this discussion of free-surface multiples. This type of multiple describes a short-path multiple that produces a series of reflections in the shallow subsurface which interact with the sea surface and shallow layers as shown in Fig. 4.6b.

Deconvolution is a commonly used method to deal with reverberations and short period multiples in shallow water. The process compresses the wavelet present in the recorded seismic data and attenuates the reverberation and short-period multiple effect, with the additional effect of increasing temporal resolution (Yilmaz 2001). A hard contrasting water-bottom reflector combined with irregular geometries can make the deconvolution process less effective, so variations of the deconvolution method can be made to overcome this (Lokshtanov 1999).

Surface-related multiple elimination (SRME) (Fig. 4.7) is a technique that predicts multiples by extrapolating the data for a single ray path journey through the subsurface and uses the data itself as the extrapolation operator to subtract multiples from the data. The algorithm uses this method to predict all the possible multiples in the

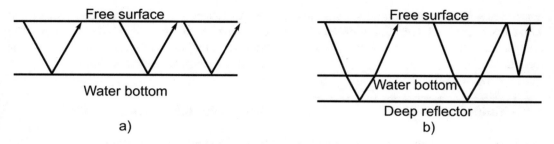

a) b)

Fig. 4.6 Diagram showing the ray paths of: **a** a free surface multiple; and **b** free surface multiple and peg-leg

data. SRME can be run as a two- or three-dimensional workflow, though consideration should be given if the multiples are in the crossline dip as the two-dimensional SRME may not be able to accurately predict the multiple. SRME has an advantage over deconvolution in that no prior knowledge of the subsurface is required while accounting for the full complexity of the Earth. Further reading on SRME can be found in van Borselen et al. (2005) and van Groenestijn et al. (2012). For further reading on techniques for water-layer reverberation removal from multicomponent data using wavefield separation see Osen et al. (1999; Osen and Amundsen 2001).

For the interpreter the range of possible effects on the data caused by water-layer reverberations are wide ranging: masking of weaker primary amplitudes, false seismic events, and general reduction of signal-to-noise ratio of the data. As with all multiple removal processes the

interpreter should assess the impact on the interval or reflector and amplitude of interest to understand the impact on the final image.

4.2.9 Incoherent Noise

Incoherent noise is highly variable both in its spatial distribution from trace-to-trace and in its amplitude range. Like coherent noise, incoherent noise can mask primary events and has a negative effect on the fidelity of the seismic image. Conversely, it is much harder to model and subtract incoherent noise from the data.

The causes of incoherent noise can be difficult to identify due to its random and inconsistent appearance on the seismic traces. Incoherent noise can be generated by natural sources such as wind, sea state, animals or man-made activity. Onshore seismic data acquired near populated areas can be particularly susceptible to random

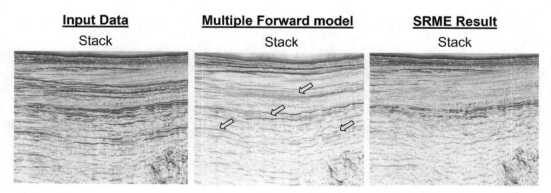

Fig. 4.7 (Left) Input seismic data with multiples, (Middle) SRME forward model identifying multiples, (Right) SRME result with multiple model removed. (Images courtesy of CGG Earth Data)

noise events from human activities: construction work, machinery and even heavy objects being dropped can all generate noise in the seismic record.

4.3 Multiples

Multiples cause problems in seismic data because they interfere with the primary reflections we want to record. Multiples are still seismic reflections, however the ray paths consist of several down and up reflections, or bounces, from multiple boundary reflections within the subsurface. This means a multiple reflection is not a true representation of the reflection coefficient of a single geological boundary. Two main categories of multiple exist: long- and short-period multiples.

A long-period multiple arrives at the receiver after the primary wave from the same shot and it is recorded as a discrete reflection event. These waves travel through more geology which reduces their amplitude content compared with primary reflections. Long-period multiples, with their longer travel time and more varied waves, are less likely to be misinterpreted as a primary event; however, they can still interfere with the primary energy from other shots.

Short-period multiples arrive shortly after a primary wave and combine with its recorded reflection. Short-period lengths have only

slightly longer travel times than the primaries and can have a similar amplitude strength. The pathway a multiple takes can involve many different ray paths, as shown in Fig. 4.8.

4.3.1 Multiples Interpretation

It is important that the interpreter is aware of multiples and has basic visual observation guides to help identify those present in a migrated seismic volume. Due to their strong and coherent amplitudes, multiples can easily be mistaken for primary energy and incorrectly included in the interpretation, especially in poor signal-to-noise data.

Some multiples can be identified based on their travel times from large velocity contrast boundaries (e.g., sea floor, geological unconformities). For example, the travel time of a water bottom multiple can commonly be identified at numerous places in a seismic section which matches multiple the TWT values from the sea surface to the seabed (Fig. 4.9). Another common way to identify a multiple is by its polarity which can be opposite and in contrast to what is expected from the surrounding reflections. The difference in polarity is due to the additional reflections made by the ray path as it passes through the subsurface. Multiples often appear to mimic the shape of the last layers they reflected off before being recorded. Where a multiple is

Fig. 4.8 Examples of multiple travel paths in the subsurface

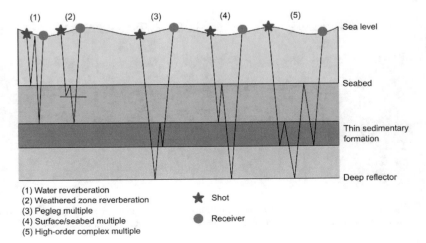

(1) Water reverberation
(2) Weathered zone reverberation
(3) Pegleg multiple
(4) Surface/seabed multiple
(5) High-order complex multiple

★ Shot
● Receiver

Fig. 4.9 Marine stacked data showing examples of primaries and seafloor multiples

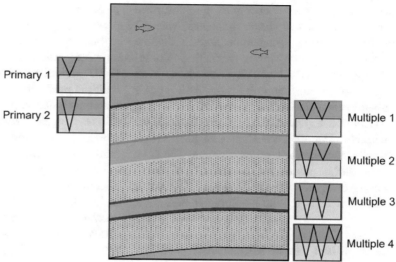

suspected the interpreter should look above and below to find identically shaped boundaries. Multiples also stick out in a seismic section by crosscutting other reflectors and not matching dips or geometries of surrounding reflectors originating from true geological interfaces. Multiples are often more obvious in post-migration gathers rather than a final stacked image, especially if they have strong parabolic moveout.

Multiples tend to show exaggerated dips compared to primary events when the subsurface has dipping geology. This exaggeration is down to the increased travel time recorded for the multiple which when migrated pushes the reflection boundary deeper into the subsurface and accentuates the dip in doing so (Fig. 4.10). When trying to image steeply dipping, complex, three-dimensional structures (e.g., salt diapirs)

having good de-multiple is very important. Residual multiple energy left in the data can be interpretated to give the incorrect geometry of a feature once migrated. Examples of this can be seen when comparing vintage and modern images of salt diapirs in the Gulf of Mexico and the North Sea, where the shape of salt structures have clearly evolved with advances in de-multiple algorithms over time. These advances are a combination of improvements in de-multiple techniques along with multi-azimuthal imaging, better velocity models and migration algorithms amplitude periodicity and normal moveout behaviour.

How velocities behave for primaries and multiples is another feature that can be used to help separate the two signal types when looking at gathers. Generally, the velocity of the sub-surface tends to increase with depth. Primary ray

Fig. 4.10 Schematic representation of the effect of dipping reflectors on multiple ray path travel times compared with primaries

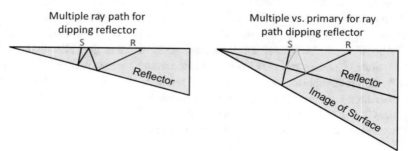

paths that penetrate deeper into the subsurface encounter rocks with faster velocities, whereas multiple ray paths spend more time in shallow layers which have slower velocities. Consequently, when NMO velocity corrections have been applied the multiples may still appear unflatten compared to the primaries.

Using these general observation rules helps the interpreter identify anomalous reflection events that may warrant further investigation. The hope is that during processing the majority of strong multiple energy is attenuated to such a degree that it does not interfere with the interpretation.

4.3.2 Multiples in Migration

Migration is a major step in the processing of seismic data. Conventional views are that migration algorithms see all the input data as primary signals and any multiples are noise that the migration is likely to incorrectly place in the subsurface. This is a reasonable assumption because the characteristics for identifying multiples rely on them appearing 'out of place' compared to the correctly migrated primary reflections. However, the multiple is still energy from the source which has travelled through the subsurface and so contains information about reflection boundaries. Therefore, multiples can be as valuable as primary reflections if the complex ray path can be understood. The clear challenge here is the ability to either model or predict the ray paths so they are correctly positioned during migration.

The different approaches to using multiples in the migration workflow can be broadly separated into techniques that aim to classify multiples with the primaries for migration and those that aim to separate primaries and multiples before migrating them as a combined data set. Similar to noise removal, transforms are an important tool to separate primaries and multiples. Given the range of different approaches not all are used in mainstream processing. The following discussion aims to provide the geoscientist with a brief overview of a variety of processing techniques pertaining to the use of multiples in modern processing projects.

Starting with methods primarily designed for multiple removal Berkhout and Verschuur (2003) describe adapting a feedback model for multiple prediction and removal using a weighted convolution process. The weighted convolution is changed to a weighted correlation process which acts as a focal transform that places primaries at zero time in the focal domain, but also converts first-order multiples into primaries. This effectively turns seismic noise into useable signal and the new data set migration is then done using conventional methods. Brown and Guitton (2005) describe a workflow that separates the primaries and multiples to obtain the maximum benefits of increased signal quality from the joint migration of primaries and multiples. Although signal quality can be improved by combining primaries and multiples using a single domain transform, serious compatibility issues exist due to ray path differences. The different ray paths for multiples require correction for the additional reflections to make them compatible for migration with primaries. Furthermore, just as multiples represent noise on primary events, the reverse is true for primaries and high-order multiples on the first-order multiple images, often referred to as crosstalk. Having separate data sets for primaries and multiples then combining them during least square migration solves these issues.

The application of reverse time migration (RTM) for the inclusion of multiples into migration is currently an area of active research. Migration algorithms which do not perform any type of complex ray path estimation, such as Kirchhoff, are less suitable for the inclusion of multiples simply because of the ray path challenges they present. RTM, which is considered a more accurate and complete solution, goes further than Kirchhoff in understanding complex ray paths and is able to predict the paths of down going and up going waves to identify and position reflection events. Liu et al. (2011) describe an approach that does not transform multiples into primaries but uses a modified RTM workflow to 'image the multiples' reflections in their correct position'. The method then migrates both types of reflection information leading to improved illumination for a subsalt case. Like the workflows of Berkhout and Brown, the multiples are firstly separated using conventional de-multiple techniques, such as Radon transform, predictive convolution or SRME, before putting the data into migration. A useful resource summary of the de-multiples and pitfalls of using multiples in RTM is discussed in Yang et al. (2015).

Certain acquisition styles favour the separation of the multiples from the primaries for inclusion in the migration. Muijs et al. (2007) make the distinction between conventional streamer and multicomponent ocean-bed acquisition, which captures up going and down going P- and S-waves. The method focuses on surface-related multiples that are mostly characterized by down going waves from the seafloor rather than up going primary waves. Separating the up going and down going waves and then carrying out 'a one-way downward extrapolation of both up going and down going wavefields' allows free-surface multiples and primaries to be used for a joint migration, which has the advantage of increasing the spatial extent of imaging (Muijs et al. 2005).

As these methods become mainstream, the energy industry is approaching the use of multiples for imaging purposes in the application of acoustic full-waveform inversion (FWI) (Sect. 9. 1). The FWI approach allows the incorporation of all type of waves regardless of the different ray paths. Presently there is no standard preference as to whether to incorporate or exclude multiples for FWI. Project time and cost is likely to play a part in the decision along with the ability to deal with issues around crosstalk (Liu et al. 2020).

4.3.3 Well-Logs

The use of well-logs to help identify multiples is another simple but effective tool. Well-logs can be used to generate synthetic seismograms that are compared to seismic traces at and adjacent to the well. Observing the match between the synthetic and seismic data may reveal additional events in the seismic that are not present in the well synthetic data. If additional events are seen to be laterally consistent in the seismic data, especially mirroring other boundaries above and with a strong amplitude, it is possible they are multiple energy. This can be taken further by introducing multiples into well-based synthetics (Luo et al. 2011). Multiple generators identified using well data can be used in broader de-multiple removal workflows using forward modelling to produce primary reflections, multiples and combinations of both with either well-logs or even VSP data. This well based workflow can be particularly useful for identifying interbed multiple generators which are often the cause of major issues for seismic imaging (Ras et al. 2012).

Additionally, well-logs can help verify the expected AVO signal in the seismic data (Chap. 6), whether it be that AVO is expected but not seen, or vice versa. In turn, this can help identify whether amplitude anomalies may be an artifact caused by multiples. One-dimensional layer modelling can also be used based on Vp/Vs ratio and density interval averages from well data to get an idea of the AVO behaviour expected from the seismic data.

4.3.4 Multiples and AVO

The presence of multiples can affect seismic based attributes and amplitude measurements taken from data beyond zero offsets and angles, such as those associated with AVO and AVA. Multiple energy often exhibits parabolic moveout when viewed on individual CMP gathers. This spatial variability across offsets and angles can lead to the multiples overlaying primary events and constructively interfering with true amplitudes, giving rise to a false apparent amplitude change. Multiples may also work to distort a true AVO signal in the data by interfering with the lateral continuity of the amplitude of a primary event, making it appear broken up and non-continuous across offsets and angles. Parabolic

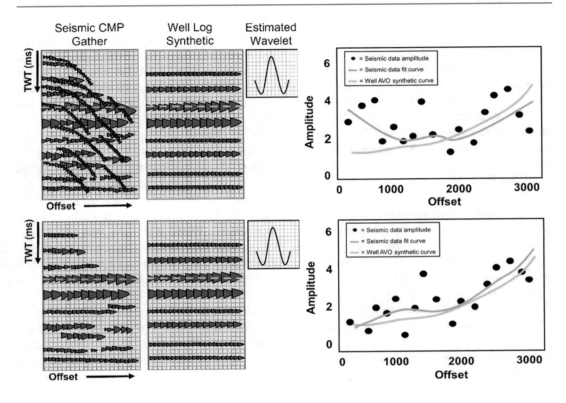

Fig. 4.11 Diagram showing the effect of multiples on seismic amplitude calculation for identifying and calibrating AVO with well data

and linear multiples can crosscut primary events across all offsets or angles. This will interfere with any AVO attributes using gradient measurement as well as amplitude maps from partial anlge stacks. This multiple-induced inconsistent amplitude behaviour can reduce the interpreter's confidence in the AVO signal. Therefore, most projects focused on AVO features, particularly on legacy or regional data sets, apply some level of post-migration data conditioning workflow. Figure 4.11 shows the effect that multiples can have on AVO curves calculated on CMP gathers.

4.3.5   Multiples Removal Techniques

Multiples attenuation, or de-multiple, in processing workflows can be applied to the data several times during processing, both before and after migration. Processing geophysicists tend to

be less aggressive in removing multiples before migration and more aggressive after because if key reflection data are missing from the migrated image due to overly aggressive de-multiple, both the de-multiple and migration will need to be re-run, which is expensive.

The methods for multiples attenuation can be categorized by the properties the techniques exploit. Some methods use spatial behaviour to differentiate multiples from primaries. We know the multiples take different ray paths through the subsurface so it stands to reason they will not occupy the exact same spatial locations in the data. An example of spatial techniques is moveout differences observed in CMP gathers. Further spatial differences between multiples and primaries can be exploited for de-multiple in post-migrations stack images such as varying dips and crosscutting behaviour. All these differences can be used as the basis for designing transforms to separate multiples from primaries followed by a filter to remove them.

Multiples in the subsurface might be attenuated following a second type of approach based on periodicity and predictions. The multiples and primaries originate from the same source, so identifying the primaries' ray paths in the data can help with predicting that of the multiples. The wave equation is commonly used for multiple prediction techniques where they be purely model-driven or data-driven.

Multiples Removal Through Differential Moveout

Primary and multiple reflections, which can have very similar travel times, are likely to have penetrated the subsurface to different depths and travelled through rocks with different velocity properties in the subsurface. A multiple does not penetrate as deeply as the associated primary from the same shot. Therefore, in a CMP gather displayed in the offset domain the visual appearance and spatial behaviour of the primaries and multiples will be very different. The application of NMO velocity correction should ideally make the primary events flat leaving the multiples to show a non-flat parabolic behaviour

(Fig. 4.12). The parabolic Radon transform filter (Hampson 1986) uses NMO corrected gathers as an input. Primaries are expected to be flat from the NMO and the multiples are uncorrected for. The Radon transform maps the data to a domain of time and curvature and, in doing so, separates the reflections into discrete data points. Due to the different curvature characteristics between the primaries and multiples the reflections are spatially separated. This then allows the multiples to be discreetly captured and removed. The data are transformed back using an inverse of the transform without the multiple energy. It is, however, possible that some multiples will have very small amounts of parabolic moveout with offset and will therefore overlap with the flattened primaries in time-curvature space. This is particularly challenging at near offset and in data sets with short offsets. The danger is that applying a very small moveout tolerance to a parabolic Radon transform will result in removing primary events that are perfectly flat.

This problem is partially solved by applying a high-resolution Radon transform (Sacchi and Ulrych 1995), which improves the resolution of the transform to increase its ability to isolate

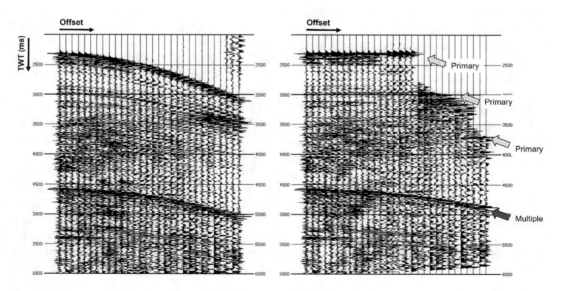

Fig. 4.12 Shows a CMP gather with multiples before and after NMO correction

primary reflections from the multiple reflections. Transform points look sharper in the time-curvature space, with less stretching. The increased resolution is achieved by extrapolating the effective range of offsets in the original data and employing a more accurate method for reconstruction of the data when transformed between domains.

Periodic Multiples Identification

Periodic multiples attenuation and removal techniques work on the principle that multiples appear at re-occurring intervals of TWT in the seismic image. This behaviour occurs because the multiple, in a simple case, bounces down and up between the same two reflection surfaces multiple times, multiplying the travel time with each up-down journey. It is therefore simple to theoretically predict where the multiple will appear in the data. For a water bottom multiple where we assume a zero-offset ray path, the first down and up wave trip to the sea surface gives a travel time. Assumptions can be made about the amplitude value of the original down going wave and the reflection coefficient of the sea surface to estimate the level of amplitude decay the multiple will experience with each trip from sea surface to floor. We now understand the period of the water bottom multiple, as well as an estimate of the multiple amplitude decay with each trip. This information can be used to predict possible periodic positions of water bottom multiples and their likely amplitude strength. As expected, the more bounces the water bottom multiple makes the deeper it will appear in TWT and the weaker its amplitude. This concept of periodic predictability becomes a more complex when we consider interbed multiples which bounce off a range of reflection surfaces at varying depths.

Adaptive Subtraction

Model subtraction involves the creation of a synthetic model that represents only the multiple energy in the data without any primaries. The wave equation can be used to predict the likely ray path of multiples. Information from the survey acquisition geometry and positioning information about the source and receiver is used along with water depth data recorded during the acquisition. The initial prediction of the multiples requires a robust initial model as it will facilitate the matching filter in the next step. The minimum energy adaptive subtraction approach is a common criterium for adaption where the aim is energy minimization of the data after subtraction (Nekut and Verschuur 1998). The multiple model is adaptively subtracted from the real data using a matching filter that removes the multiple energy whilst preserving the primaries. A matching filter is required because the initial multiple energy model is likely to have imperfect estimates of multiple amplitudes, phase and frequencies. Simply subtracting the initial multiple model from the data would most certainly damage the primaries. The matching filter improves the fit of the initial multiple energy model to real data and then tries to generate a more accurate estimation of the multiples by an adaptive matching technique. The estimation and design of the matching filters is just as important as the initial multiple model, as any errors occurring here can damage the effectiveness of the workflow.

The process of adaptive filtering is not straightforward because the presence of signal and other noise in the data interferes with the matching process. Primaries damage is a risk when applying this method and the interpreter should observe the final effects on a fast-track migrated stack during the active processing workflow. Abma et al. (2005) gives an introductory comparison of various adaptive subtraction methods.

Surface related multiple attenuation (SRME) is a predictive and subtractive data-driven method that deals with free surface multiples. The method can predict not only water bottom multiples but also peg-legs and diffracted multiples. The two-stage approach can be thought of as a form of the adaptive subtraction method in that, firstly, multiples are predicted by modelling from the primary ray paths using all the starting points for down-going multiples. And, secondly, adaptive subtraction is used to remove the predicted multiples. The methodology of SRME prediction can also be applied to internal multiples, however the application is focused on a single layer interval. The work of Verschuur (1992) provide a good introduction to the foundations of the SRME technique and its application.

4.3.6 Interbed Multiples

Interbed multiple attenuation (IMA) is a notably more difficult method in a seismic processing workflow when compared to free surface multiples. Interbed multiples are difficult to differentiate based on velocity moveout differences because they are often reflected on very closely spaced events, producing a very similar ray path and velocity to the primaries. Consequently, Radon based methods are not as effective. Interbed multiples also have very similar amplitudes and frequencies which reduces the usefulness of predictive filtering. In addition, the multiples dips are similar to the primaries and, thus, reduce the effectiveness of approaches like F-k filtering, Tau-P de-multiple and migration methods that use dip discrimination techniques

during imaging (Griffiths et al. 2011). When identifying multiple generators (i.e., a surface that causes both upward and downward multiple ray paths) they are mainly associated with events in the seismic data with large reflection coefficients. Interbed multiples may be generated from boundaries with weaker reflection coefficients, but they will likely produce negligible amplitudes (Brookes 2011).

As discussed earlier some multiples can be identified using periodicity which can easily be measured from obvious large velocity contrasts between two layers. However, interbed multiples ray paths must make a minimum of at least three reflections before being recorded at the receiver, and so periodicity can also be used to identify a range of reflection coefficients throughout the subsurface off which multiples may have bounced. These velocity contrast features are referred to as multiple generators. Typical multiple generators are identified within a migrated seismic section where they exhibit strong amplitude, for example unconformity surfaces or where a distinct change in geology occurs.

The spatial pattern of multiples can be used to distinguish them from primary events in the de-multiple technique known as pattern recognition (Spitz 1999) (Fig. 4.13). The technique models specific multiple using seismic interpreted

Fig. 4.13 (Top image) Input seismic stack with interbed multiples highlighted. (Bottom image) Output seismic stack with multiples removed. (Courtesy of CGG Earth Data)

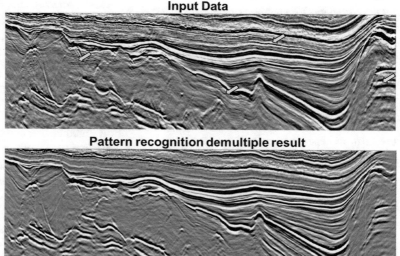

Input Data

Pattern recognition demultiple result

horizons as multiple generating surfaces allowing prediction of their shape and time and is often used to specifically target pegleg multiples often referred to as SPLAT (Specified PegLeg Attenuation). The techniques can be quite successful however it has the limitation of only having the interpretation to use as multiple generators (Mojesky et al. 2013). If the seismic is challenging to interpret this technique is harder to use.

Data-Driven and Model-Driven Techniques

There are two main recognised approaches to dealing with interbed multiples: model- and data-driven approaches. Model-driven methods go beyond using just seismic reflection data to differentiate between primary and multiple energy and, in fact, attempt to model the expected internal multiple based on wave theory. These techniques can include building three-dimensional models to include several multiple generators picked from the seismic data, along with additional prior knowledge such as a velocity model. Such models can be used to help predictively model interbed ray paths (i.e., propagation and back-propagation). Good results rely on accurate information about the subsurface and an overly simplified model is unlikely to predict complex interbed multiples. Many model-driven methods are a hybrid that combine aspects of both data- and model-driven techniques. In data-driven approaches different primary reflection events present in the recorded data are used to predict various possible interbed ray path combinations using convolution and cross-correlation methods (Ma et al. 2020). For the results to be accurate and meaningful these techniques rely on the signal-to-noise quality of the data, as well as input from the interpreter in the form of seismic horizons.

The geoscientist should be aware that the quality of the input seismic data within the interval of interest will play an important part in defining the limitations for successful IMA using data-driven methods. Even modern input data can exhibit areas, especially deep in the subsurface, with low signal-to-noise, poorly illuminated complex geology and limited frequency content which will not respond well to even the most involved IMA workflow. Applying IMA too aggressively in areas with limited data quality is likely to damage the primaries, thus reducing the interpreter's ability to map important features. Processors will often attempt to compensate for overly aggressive IMA with Q-amplitude processes, but this approach cannot restore reflectively that has been completely removed from the data.

An industry standard approach for data-driven interbed de-multiple is described by Jakubowicz (1998) as a data-driven predictive method using the wave equation based on a single generator surface picked from the data. There is no requirement for earth model generation as the method focuses on a window of data for the prediction, rather than the whole data set. Consequently, the method is limited by a layer-by-layer approach, known as layer stripping, that is used to account for different multiple generators. The process of application is therefore slow, computationally expensive and impractical on large, modern three-dimensional data sets (Dutta et al. 2019). The accurate interpretation of the multiple generators is key, and the interpreter should ensure that any surfaces used are properly QC'd as errors can cause a cumulative error throughout the section.

Inverse scattering series (ISS) is another data-driven de-multiple technique used for IMA. Picked horizons are required to act as multiple generators to predict all possible internal multiples (i.e., diffraction, converted waves, head waves, diving waves) for all possible generators at once. This algorithm is highly useful for the interpreter when the subsurface is complex and the interpretation of potential large velocities boundaries is too challenging. The ISS attenuation predicts the correct travel times and approximate amplitudes of all the internal multiples in the data including the converted wave internal multiples. The method does require the seismic data to be de-ghosted and have all free surface multiples removed (Fu et al. 2010). ISS is particularly useful in frontier exploration areas or areas of complex subsurface geology where challenges exist with interpreting multiple generators or reliable seismic velocities. One of the clear drawbacks, however, is its high computational cost. For further reading on the ISS technique see publications by Weglein et al. (1981, 2006) and Matson et al. (1999).

The Marchenko method, like ISS, does not need surface interpretation of the data or require layer-stripping to deal with generators separately. The technique uses up and down focusing functions by solving Marchenko equations. This allows for the retrieval of Green's function (Thorbecke et al. 2021) which describes motion induced by an impulsive point source to generate all orders of multiples as well as primaries at any depth with no requirement to solve the shallower multiples first (Staring and Wapenaar 2020).

Ikelle (2005) describes another data-driven method that uses 'virtual events' to represent multiple generators that are predicted but not recorded. Virtual events are used in conjunction with recorded events to predict scattering of internal multiples, similar to those from a free surface. Here, the data are separated into 'segments' to control the range of multiple periods generated. It could be argued that this method is not strictly data driven and is more of a hybrid that includes modelling elements.

A commonly referenced model-driven technique using wave extrapolation is described by Pica and Delmas (2008). The method follows a similar workflow to data-driven techniques but as a model-driven method it requires the input of both a velocity model and reflectivity model. Verschuur and Berkhout (2005) describe the application of a technique that uses input of a velocity model and extends on techniques used for surface multiples attenuation where shot records are replaced with common-focus-point (CFP) gathers, which aim to deal with the downward-scattering. Berkhout and Verschuur (2005) affirm the algorithms can be applied in a model-driven or data-driven way, though we recognize it here as a model-driven method.

One final technique worth noting is based on the application of inverse theory for the removal of interbed multiples. Ramirez and Weglein (2008) and Weglein (2009) discuss the merits of linear inversion as a tool for de-multiple, which reaches beyond the conventional wisdom and ubiquitous use of seismic inversion as an exploration tool (Chap. 6). Again, these techniques sit somewhere between data-driven and model-driven.

4.4 Velocity Models in Seismic Processing

Throughout the seismic processing workflow estimations of the velocity variations of the subsurface are required to help produce a final seismic image suitable for interpretation. Velocity models are also used to convert seismic data from time-to-depth domain and back again. The building of velocity models can be simple or complex, which is driven by the complexity of the subsurface geology. To produce a velocity model the subsurface is, essentially, reduced to layers with different velocities assigned to them, although special geobody features such as salt diapirs go beyond this simple model. Velocity models have many different applications depending on the stage in the seismic processing flow. Generally, the later in the flow the more complex a velocity model needs to be. The final velocity model should always try to be aligned with and related to the geology. There is often a disconnect between a final velocity model, which from a processing perspective flattens CMP gather events, yet from a geological perspective the variability and values of velocities do not match the geological understanding. The resolution and level of detail in a velocity model also needs to be considered appropriate for its application. A coarsely picked residual moveout (RMO) model is suitable for gather flattening, but for a complex RTM within structurally complex geology it would not be appropriate. Jones (2018) reviews the main concepts regarding seismic velocities and velocity model building.

One common pitfall of velocity model building is to create an overly complex model, which though technically superior fails to deliver a significant uplift of the final image over a less complex but more time effective model. For reprocessing projects there can be a tendency to discard an existing velocity model as being inferior only to reproduce something very similar at the end of the project. The building of velocity models and their application has been extensively discussed (e.g., Ylmaz 2001). In this book we instead look at data sources for supporting velocity model workflows and highlight any pitfalls from an interpreter's perspective.

4.4.1 Well Data, Check Shots and VSPs

The only ground truth measured data from the subsurface available to the geoscientist comes from wells, and as such well data are a valuable source of velocity information for model building. Virtually all drilled wells, regardless of vintage, will acquire P-sonic logs with most modern wells also running S-sonic logs. Although sonic logs are very useful for identifying large velocity boundaries, for velocity model building they only provide limited measurements into the formation from the sidewall of the borehole. Sonic logs are also susceptible to borehole conditions (e.g., washed-out areas and fluid invasion) and measurement errors (e.g., cycle skipping), and sonic tools can struggle to record meaningful data in very low-velocity formations.

Wells also have other sources of velocity measurements in the form of checkshot surveys or vertical seismic profiles (VSP). VSP surveys range in complexity from a simple string of geophones placed at intervals in the borehole with a checkshot, to full walkaway surveys providing a high-resolution seismic survey around the borehole (Fig. 5.5). Checkshot surveys are sensitive to a much larger area of the subsurface around the borehole than a sonic log and give a good indication of velocities between boundaries known as interval velocities. The checkshot survey is often used to correct the sonic log when well tying is done to correct any velocity inconsistencies (Chap. 5).

VSP surveys are considered the preferred borehole seismic survey and offer the ability to obtain high-resolution seismic images around the borehole. There is a vast variation of source receiver configurations for VSP surveys and receivers are generally spaced more closely than a standard checkshot survey, which provides more detail even in the most basic configuration. VSPs use hydrophones, geophones and accelerometers to record direct arrivals, but also up going and down going waves, as well as multiples. Three-component VSP receivers can also record both for P- and S-waves.

The geoscientist should consider this variety of well-based velocity information as a useful ground truth when either building a velocity model or evaluating an existing velocity model. Like all data it is advisable to get data quality confirmed by a specialist particularly because VSP data quality issues might not be as evident as in a P-sonic log data, which shows clear borehole effects. Although these well-based data are useful, pitfalls using wells that show large velocity discrepancies within the same formation compared with other wells should be carefully investigated. Depth trends, over pressure, localized geological events (cementing, erosion, reworking) (Sect. 6.5) and structural history can all cause velocity variability in wells, so the geoscientist should try to broadly understand these with the help of specialists from other disciplines.

4.4.2 Velocities for Migration

Velocity model are an important input for the migration workflow. The velocity field is used to determine the ray paths of the recorded reflections and represents the wave velocity for each subsurface point. In addition, the velocity field allows the migration to move reflectors into their correct subsurface position.

For the interpreter the choice of migration algorithm is important because it affects how reflection events appear in the final image. The processing geophysicist will play a key role in selecting the appropriate migration algorithm. Factors such as magnitude of geological dips, spatial variability of geology spatially and structural complexities are just a few of the many considerations. Factors of importance to the interpreter are the level of velocity control (i.e., wells, checkshot, VSP) and geological input (horizons) that are available to build an appropriate velocity model for migration algorithm of choice, compared to what is ideally required by the processing geophysicist.

Post-stack and Pre-stack Migration Velocities

One of the many choices when choosing a migration method is whether to use pre-stack or post-stack input data, which will influence some aspects of building the velocity model. Post-stack time migration takes an input of CMP traces that are corrected for NMO and then stacked together creating what is assumed to be normal-incident reflections. This is a very broad assumption and is certainly not appropriate for surveys with long offset or complex dipping geologies. An advantage of the post-stack migration is that there are less traces to migrate and the process is, therefore, much quicker compared to the pre-stack method.

Most modern processing flows favour pre-stack migration where traces are arranged in common offset gathers rather than by CMP gathers or stacks. This method has a range of imaging benefits such as a better ability to deal with dips and improved image resolution. Carrying out pre-stack migration also avoids stacking conflict which arises when two events at different TWT occupy the same place on the CMP gather but require different velocities to correctly flatten them.

So, what does the migration choice mean for the velocity field? For post-stack migration the velocity model variation is in the vertical direction and is not required to accurately capture horizontal velocity variation, as the migration is focused on straight ray paths approximations.

Pre-stack migration is designed to handle more of the lateral velocity variations, which means the velocity model must account for areas of thickness change, compaction, over and under pressure along with some constraints based on the geology. Therefore, the velocity requirements for pre-stack migration will need more spatial detail and control from well data to get the most from the migration.

4.4.3 Stacking Velocities

Stacking velocities are a common source of velocity information for the geoscientist. They are used in the NMO correction step of seismic processing. Stacking velocities are used to correct for the hyperbolic movement of the primary events seen in the CMP gathers, which are the result of the travel time changes as a function of increasing source receiver offset. This velocity field is referred to as stacking velocities and produces what are known as average velocities. The traces are then summed or stacked together in a process called CMP stacking. Rather than using average velocities the geoscientist often prefers to convert them to interval velocities as they give a more representative picture on the velocities. A common method to convert average velocities to interval velocities is the Dix equation (Dix 1955):

$$v_n = \sqrt{\frac{V_n^2 \tau_n - V_{n-1}^2 \tau_{n-1}}{\tau_n - \tau_{n-1}}}, \qquad (4.5)$$

where v_n is the isotropic velocity within a layer, n is the layer number, τ_n the TWT and V_n are the RMS velocities.

When comparing average velocities of a theoretical model to the converted interval velocities we can see that the interval velocities are useful for identifying the boundaries of layers, but do not capture the velocity variability within the layer. Stacking velocities are often automatically picked using software, however, they need to be checked and it is important to be aware of the density of the picking that has been carried out. Where the subsurface geology shows smaller localized changes a coarse stacking velocity picking interval may miss these nuances, leaving undercorrected gathers. Slopes seen within the semblance volume used for velocity picking should be accurately captured to avoid producing visual artifact effects if the average velocities are converted to interval velocities. Stacking velocities picked at different levels of detail can produce gathers that appear equally flat when the velocity behaviour of the subsurface is broadly uniform for the given geological layers. However, the detail in the stacking velocities can become very important when areas of localized velocity behaviour changes are present in the subsurface and the data. Automated picking may not be sufficient in these cases and some manual intervention by the processor and geoscientist may be required. The ability to pick detail depends on the quality of the CMP data. Densely picked velocities might appear noisy and it is common practice to apply some level of smoothing to remove sharp edges.

4.4.4   Average Velocities

If we assume the case of a vertical ray path coming down from our source with non-dipping layers in the subsurface, then we can work out the velocity of the rock down to the top of that reflection event by simply dividing the depth to the event by the one-way travel time. This is known as the average velocity. The average velocity value calculated at one position can be very different from somewhere else. Variations in layer thickness over the survey and changes in geology will have an effect. However, it gives us a basic understanding of velocity.

4.4.5   Root-Mean-Square Velocities

The root-mean-square (RMS) velocities are suitable to use when there are multiple horizon layers. Each layer will represent a discrete

interval velocity. The RMS velocity value represents the square root of all the layer velocities squared divided by the number of layers:

$$V_{RMS} = \sqrt{\frac{v_1^2 t_1 + v_2^2 t_2 + \ldots + v_1^2 t_N}{t_1 + \ldots\ldots + t_N}}. \qquad (4.6)$$

The RMS velocities will be lower than interval velocities of the layers. The offsets should ideally be small $\sim$2–3 km and isotropic conditions assumed in order to be a valid representation. Often these velocities are described as being equivalent to NMO velocities.

4.4.6   Interval Velocities

Interval velocities (V_{int}) are sometimes called layer velocities and represent the velocity value of the seismic energy passing through a single layer. Interval velocities can also be estimated using sonic log measurement and even laboratory measures made through borehole core. It is worth considering that several smaller layers may comprise a bigger layer; if this is the case then this velocity might be considered close to an average velocity. Interval velocities can be calculated using the Dix equation (Eq. 4.5). The equation uses V_{RMS} velocities and assumes flat parallel layers that are isotropic and have limited offset:

Figure 4.14 shows a theoretical relationship between average velocity, V_{RMS} and V_{int}.

4.5   Time-To-Depth Conversion

Seismic reflection surveys record data in TWT that represents the ray's path from the source down to a reflection event and up to the receiver. More than ever, processing projects use depth migration because it gives better image quality and more accurately recovers reflector geometries compared to time migration. Another motivation for using depth migration is that time domain is of little practical use for many of the tasks the geoscientist and engineering disciplines

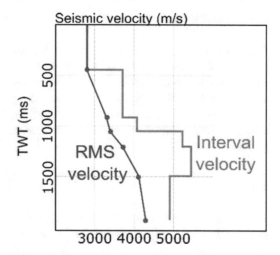

Fig. 4.14 Schematic representation of the relationship between seismic velocities

require, such as depths or thickness measurement for well planning, volumetrics, column height estimations.

However, seismic data are not always depth migrated and, so, the conversion of seismic data and interpretation from time-to-depth using a suitable velocity model is a particularly important geophysical process that is applied when assessing the structural and depositional evolution of a broad regional areas, or the finer scale of prospect evaluation for well placement planning.

A pre-stack depth migration (PSDM) volume coming out of migration should not be considered as the final answer for a seismic volume in depth, as there are likely to be areas where the migration velocities are over simplified or have an incorrect velocity for a particular formation, which will require additional work to refine the velocity model. Having seismic data in the time domain is preferred for the task of seismic well ties as it requires a constant wavelet for the convolutional process of generating a synthetic seismogram, which is difficult to achieve in depth domain deconvolution (Singh, 2012) (Chap. 5). Geoscientists may want to convert depth domain seismic data to the time domain to evaluate the accuracy of depth migration velocities with seismic well ties. Once in the time domain the geoscientist may want to make refinements to the

existing migration velocities to improve any observed mis-ties. Testing alternative velocity model building techniques is always good practice. Once the geoscientist is happy with the new velocity model it can be used to convert the time domain seismic data back to the depth domain. Updating an existing velocity model used for migration is common practice by adding new subsurface information from sources such as newly drilled well data.

This section offers some information for velocity model updates assuming the seismic data are in the time domain and the interpreter wants to convert it into the depth domain. Workflows for migration velocity generation are not discussed here, although varied information can be found on processing velocities throughout Chaps. 3, 4 and 8.

4.5.1   Interpretation Considerations

Geological structures and features within the subsurface naturally exist in the depth domain, and we know that many of the measurements required to assess and operate in the subsurface are required in depth domain. Well planning and calculations required for economical assessment and risk uncertainty require volumetric calculations which would be meaningless in measurements of TWT. However, prior to dealing with these challenges it should be acknowledged that the visual appearance of the subsurface may be quite different depending on whether time or depth domain data is being used. There is therefore a direct implication as to how an interpreter might describe subsurface structures differently between domains and when this might become a problem.

Complex structural interpretation can be very challenging, even with the best seismic data, because it requires good illumination of irregular and highly dipping features, correct lateral and vertical positioning and sharp imaging. Legacy time migration workflows such as post-stack time migration (PSTM) use simple velocity models to describe the geology contrasts in the subsurface, but struggle to accurately image dips of relatively simple and complex structures. Historically, when most exploration prospects relied on purely structural trapping mechanisms, such as four-way-dip anticlines, this was a real issue for the interpreter. Structural surfaces mapped in the time domain and then converted to the depth domain could experience significant changes of structural dips known as flexing. So, what once looked like a viable trapping structure with large volumes became a much flatter poorer quality structural trap with greatly reduced volumes making it less economically appealing.

Stratigraphic traps or combination structural and stratigraphic traps are still sensitive to domain changes, perhaps more so than large structural traps which tend to exhibit tens if not hundreds of meters of trapping geology. Traps relying on faults seals can also be sensitive to domain changes both related to the orientation of the fault plan and magnitude of vertical offset required for trapping.

Although pre-stack time migrations are a better approach and velocity model building is far more advanced, doing complex structural interpretation in the time domain is considered riskier than in the depth domain. However, this is probably an oversimplification and the advice for the interpreter is to assess the quality of the available velocities and the amount of movement in reflections and structures between time and depth early on in a project.

The importance of structural conformance of amplitude anomalies is explored further in Chap. 6 but is also relevant here for the geoscientist. Amplitude maps made from time domain seismic data and displayed on interpreted surfaces picked in the time domain can show very convincing conformance to structure contours. The same amplitude maps should also be re-created using depth converted seismic data on depth converted structures to check that the contours are still closed around the anomalies and the amplitudes are not seen to spill irregularly across the structure.

4.5.2 Data Sources

The construction of a velocity model for the purpose of time-to-depth migration aims to accurately capture the variability of the velocities of the geology in the subsurface. To successfully achieve this relies on an understanding of all the different data sources of velocity information available.

Well Data

Well data as direct measurement of subsurface velocity can be used to independently produce velocity maps and even three-dimensional velocity models using seismic interpretation (i.e., horizons and faults) to delineate boundaries and define layers (i.e., guide the spatial interpolation). To produce these maps or volumes of velocity we need to overcome the spatial scarcity of the well data by using a form of interpolation method to populate the larger area of interest with velocity information from the well. Numerous interpolation methods exist such as nearest neighbour and inverse distance triangulation. Kriging is a geostatistical interpolation method used to predict properties between observation points, accounting for a spatial pattern as revealed by a spatial covariance matrix. These well-based maps or volumes can be combined with velocity maps or volumes from seismic velocities, using methods such as co-kriging, to get the advantage of information from both data sets.

Seismic velocities from processing are the other main source of velocity information and in some cases may be the only source available to the geoscientist. Velocities used for processing may not be the best velocities for time-to-depth conversion as they are often created with a specific purpose (i.e., stacking velocities for flattening gathers or migration velocities for positioning reflectors and collapsing diffractions). Velocities for time-to-depth require true vertical velocities that reflect geology (Etris et al. 2002). However, the lateral coverage of seismic velocities is usually very good but may require additional smoothing and interpolation.

4.5.3 Model Building Approaches

All Velocity models for time-to-depth conversion are constructed using a series of seismic surfaces that form layers in the model, which the interpreter has made or inherited. The surfaces in the model should match geological markers present in any wells being used in the model building. Quick time-to-depth conversion can be done using only seismic velocities by assigning an average velocity value to each surface for a single layer. This quick method can be improved by converting the average velocities to interval velocities which are able to time-to-depth convert multiple layers at once (Francis 2018). This method is considered suitable when a geological layer is laterally continuous with smaller lithological variations, where the depositional and diagenetic history is simple and steady compaction has occurred over geological time.

Incorporating well data is always preferred. Co-kriging can be used to combine well data with the seismic velocity information to produce a more geological velocity model for time-to-depth conversion. Using well-based velocity information to generate linear interval velocity depth trend functions is a common approach to velocity building. These trends can be used in the form $V_0 + K$, where V_0 is a reference velocity and K represents the gradient of the velocity change within a defined layer in the model. This is a popular approach but can be limited as we assume a linear gradient. Complexities such as velocity inversion can still be implemented using this method and just require the correct layer boundaries and velocity gradient from the well data. The geoscientist should be careful of this approach when dealing with steep gradients in thick layers or velocity values deep in the subsurface with sparse well control, as overly steep K values can lead to unnaturally fast velocities being introduced.

Time-to-Depth Scale

The scale of a time-to-depth velocity model largely controls the choice of approach and level of detail required in the model. Regional velocity models can cover tens of thousands of square kilometres and include hundreds of wells. Conversely, the scale of a prospect sized time-to-depth conversion could be less than one hundred kilometres square and contain five or less wells. The scale is also directly proportional to the acceptable depth conversion error using the velocity model. A depth error of a hundred meters on a regional scale depth conversion can easily be accepted and may only represent an overall five to ten percent error based on the total depth of the model. However, on a prospect level scale, where volumetric calculations underpin economic viability assessments, depth conversion accuracy is a lot more important down to several meters. The interpreter should aim to create the simplest velocity model for time-to-depth conversion, but it must be accurate enough for its application.

When building a large regional velocity model, the coverage of seismic velocities is not always continuous, so we must rely on well-data only based workflows. In this case it is tempting to include many wells in a bid to increase accuracy of the velocities and improve the model. However, this commonly leads to velocity artifacts in the model, as even the most common spatial interpolation techniques struggle to make sense of a combination of variables (i.e., large differences in well spatial density, large velocity variability in closely spaced wells, variable data depth extents). For smaller areas with a lot of wells the effects can be even more extreme. A best practice approach is to spend time carefully picking suitably spaced wells of similar depth extents that have good quality velocity data. Once an initial model has been generated by the chosen method more wells can be added to increase spatial variability and localized complexity. The same is true in defining layers within a velocity model for regional purposes using regional interpretation surfaces. Only a confidently picked surface should be used in the model and the surface must correspond to available well markers. Too many surfaces will likely result in velocity artifacts as a function of thickening and thinning of the formation.

4.6   Anisotropy

The subsurface is complex in places and at some scales it is organized and homogeneous whilst in others it is chaotic and heterogeneous due to the variable geological processes that have created it. As such, subsurface velocity properties would not be expected to fit the description of isotropic, by which we mean there are no subsurface spatial patterns with a preferred direction. Anisotropy is the opposite of isotropic and describes materials that have different properties in different directions, and specifically properties that effect seismic wave propagation.

Seismic anisotropy is defined by the variable seismic wave velocity dependent on the orientation of travel through the subsurface. Waves produced from a seismic survey source propagate through the subsurface at a range of angles and, as such, anisotropy should be considered when processing seismic data. Until recently subsurface velocities have routinely been treated as isotropic because historically it was perceived as too difficult to accurately account for anisotropy with technology at the time. Additionally, the effects of not accounting for anisotropy were assumed to be insignificant as they had not been linked to causing wide issues with inaccurate velocities and large depth uncertainty. However, numerous case studies have shown that the magnitude of anisotropy is important, which has led to it becoming a more frequently performed analysis and corrected for in modern processing projects. Consequently, it is important that the geoscientist is aware of causes, effects and instances where anisotropy may be more likely.

4.6.1   Causes of Anisotropy

When seismic waves travel though a homogenous sandstone layer with evenly distributed and

equally fine-sized quartz grains, it may appear to be isotropic. Shales are usually classified as being anisotropic due to high concentrations of clay minerals, which are predominantly sheet silicates. During shale deposition and burial, the sheet silicates can become aligned which increases anisotropy. Other factors such a kerogen, micro factures and micro lamination are all potential contributors. Deng et al. (2021) provide a useful modern real data analysis paper on the properties of shale. Measuring the presence of anisotropy in a particular formation indirectly tells the geoscientist something about the composition of lithology (Winterstein, 1986). A study by Kendall et al. (2007) of siliclastic rock in the UK Clair field attempts to relate findings from laboratory measurements with AVOA (amplitude variation with offset and azimuth) from seismic data to help identify areas of increased fracture density which are directly related to reservoir quality.

When assessing the subsurface for the potential of anisotropy there are several commonly documented geological features that should be considered:

(a) Horizontal geology with fine alternating layering of laminations;
(b) Stress regime induced anisotropy caused by tectonics;
(c) Presence of fracture sets;
(d) Shales.

4.6.2   Scale

The issue of scale is fundamental to the understanding of anisotropy as it is variable at a range of different scales from the very small mineral level to much larger formations. The geoscientist may assume that if the thickness of a particular geological layer (e.g., shale with anisotropy properties) is smaller than that resolvable thickness by the wavelength of seismic data, then the presence of anisotropy might not be a consideration; however, this is not the case. Layers that are unresolvable by seismic waves can still produce anisotropy effects within seismic data (Thomsen 1986). These thin unresolvable layers are often considered common contributors to the presence of anisotropy and can be verified from borehole logs that have a much finer vertical resolution compared to seismic data. This also applies to sub-seismic fractures and faults which can be difficult to detect on seismic data but can be detected on data such as borehole image logs. The variability of particles that make up the matrix or framework of a particular lithology play an important role on the anisotropy properties. Baden et al. (2017) show that the scale on which anisotropy measurements made in carbonates directly affects the amount of anisotropy recorded in laboratory samples. The study makes the important conclusion that anisotropy from the matrix can be "overlapped" by larger scale fracture induced anisotropy effects, as well as a clear lack of correlation between seismic-scale measurements and those are at laboratory scales. Therefore, it is likely that a level of anisotropy exists within different measurement scales of the same lithology.

Scale also needs to be considered on a practical level when looking at the necessity of accounting and correcting for anisotropy effects on seismic velocities. It is impractical and potentially not required to construct a velocity model that attempts to capture all the variability of anisotropy at all depths. However, if a target interval sits directly beneath 1500 m of structurally complex dipping overburden with a migration velocity uncertainty of 20% (due to anisotropy), which relates to a potential depth error of ~ 300 m (20% of 1500 m), this scale of anisotropy should be addressed. Lateral positioning errors also exist due to anisotropy velocity effects when comparing isotropic and anisotropic depth migration, and can be in the order of over 100 m (Holt and Campbell 2008).

4.6.3 Types of Anisotropy Models

Polar anisotropy is characterized by horizontally alternating strata layers which give variation in velocity and symmetry in the vertical direction. This type of anisotropy is commonly encountered by the interpreter when dealing with seismic data, simply because a high percentage of the subsurface geology of interest for energy resources is constructed of alternating layers in this orientation. The simplest way to visualize this is a simple binary sand shale system. Each of the individual sub-layers taken in isolation is transversely isotropic. As the anisotropy is observed in the vertical direction polar anisotropy it also referred to as vertical transverse isotropy (VTI). Instead of anisotropy, the isotropic nature of the strata perpendicular to the vertical polar axis is referred to. In strata identified as VTI, anisotropy changes with angle in reference to the vertical polar axis (Fig. 4.15). For VTI we need to derive three parameters (i.e., velocity, delta, epsilon) from seismic data (and wells) rather than one (i.e., velocity) for isotropic.

Geology maybe be deposited in horizontal layers however this does not mean that it stays like this. Tectonic processes and localized stress regimes can tilt previously horizontal layering. This means the axis of symmetry defined for VTI would no longer be appropriate and must be tilted to compensate. This type of anisotropy is referred to as tilted transverse isotropy (TTI) (Fig. 4.16). If a VTI velocity field is used for seismic migration with data where a TTI velocity field is more appropriate because of steeply dipping geology, then negative effects on image focusing and positioning of reflectors can occur. Guan et al. (2008) show a convincing two-dimensional real data example comparing VTI and TTI RTM for highly dipping sedimentary strata undispersed with salt diapir.

Beyond TTI, the horizontal direction can also be an axis of symmetry showing variation in velocity and is referred to as horizontal transverse isotropy (HTI). As geological depositional processes are biased to horizontal layering, areas where HTI velocities are needed will have steep dips and therefore likely have a structurally complex history (Fig. 4.17). HTI is also applicable where vertical fracture sets are observed. TTI has additional complexity in that the velocities will be isotropic in the vertical direction and

Fig. 4.15 Vertical
transversely isotropic
subsurface model

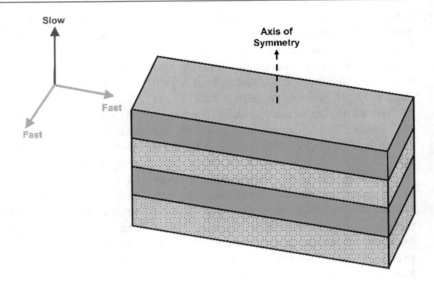

Fig. 4.16 Titled transverse
isotropy subsurface model

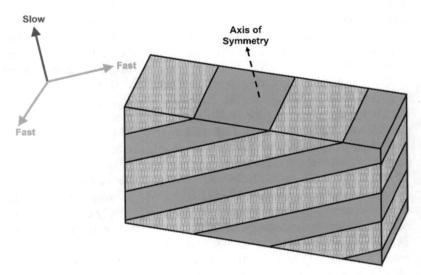

in one of the horizontal directions which does not
show symmetry. This is also referred to as azimuthal anisotropy where the rock properties vary
not only on the angle of incidence from the polar
axis of symmetry, but also the direction of
propagation through the medium. For TTI we
must derive five parameters that include the
previous three mentioned for VTI along with
those for in-line and cross-line dip.

Orthorhombic is the final model consideration
for the anisotropy medium, which is also
described as being azimuthal, and represents the
most complex variation of velocities. Unlike the

VTI, TTI and HTI a consistent uninterrupted
direction of isotropic velocities does not exist.
Instead, continuous medium is broken up by
perpendicular discontinuities creating a sugar
cube effect (Fig. 4.18). As such, three different
velocities (i.e., one vertical and two horizontal)
orthogonal to each other are observed. To fully
observe the effects of azimuthal anisotropy multi-
azimuthal acquisition (e.g., multiple narrow-
azimuth surveys, wide-azimuth survey, full azimuth survey) is required. For modern land surveys this is fairly standard but for marine surveys
it is not yet recognized as standard practice. The

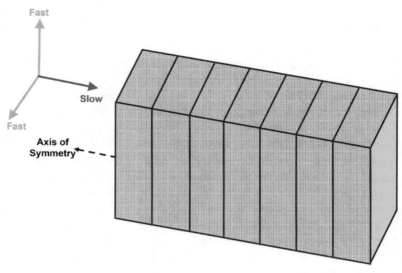

Fig. 4.17 Horizontal transverse isotropy subsurface model

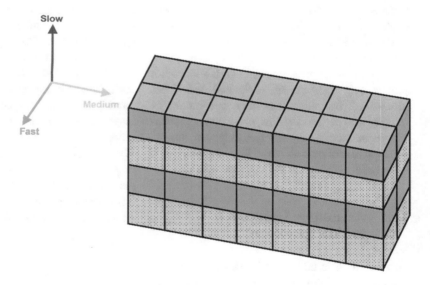

Fig. 4.18 Orthorhombic subsurface model

number of required derived parameters from seismic goes beyond the five for TTI, however depending on the processor and available algorithms the number of additional parameters can vary.

4.6.4 Thomsen Parameters

Wave propagation is controlled by elastic properties of a material that give information about how the material responds to stress and strain (Chap. 2). The elastic properties have a direct control on the velocity of seismic waves. For P- and S-waves we must consider the two elastic constants of incompressibility and rigidity, which are measured with bulk modulus (k) and shear modulus (μ). For an isotropic material we are only concerned with compressional waves and shear waves, which provide input values to calculate the required elastic constants. In introducing the concept of anisotropy, the elastic constants become dependent on the direction of wave propagation, which requires a more complex calculation. We also need to consider three wave directions comprising the P-wave and two S-waves (although technically they are not strictly polarizing and completely orthogonal to each other and can be referred to "quasi") (Liu and Martinez 2012).

Even today, the foundation for all anisotropy work is based on parameters from Thomsen (1986) who observed that most anisotropy effects on velocity are weak, in the order of 10–20%, and allow considerable simplification to be made in equations used to estimate the effects. Thomsen introduced the parameters Delta (δ) for short-offset almost vertical velocities, Epsilon (ε) for long offset effect including horizontal velocities and Gamma (γ) for shear wave effect, as well as defining three primary directions x, y, z. Each parameter comprises several elastic moduli constants that are defined using elastic tensors, which allow the elasticity to be "*represented more compactly*" (Thomsen 1986). He makes it clear that δ is the most even, though it does not include any measure of horizontal velocities. Along with velocities at normal incident for P-velocity and S-velocity this gives five independent parameters to characterize polar anisotropy. The equation allows an angle from the polar axis to be used to calculate the velocity of a ray.

4.6.5   Anisotropy Implication for Interpretation

An interpreter is interested in the effects of anisotropy on the migration of reflection events and its possible effects on AVO. The magnitude of the effect is related to the magnitude of the anisotropy present in the subsurface. Accurate pre-stack depth migration requires accurate velocities for a well-focused image. When velocities are incorrect the calculated travel times in the migration will also be incorrect, which leads to spatial mis-positioning of reflections whereby reflection events do not appear crisp, the edges of complex structures are slightly blurred or the continuity of reflectors with high dips is broken-up. The correct depth position of reflections from

migration is another reason to address velocity anisotropy. Migration depth error magnitudes of 10–20%, which could translate into hundreds of meters variability, are quite feasible. For prospect identification and well placement around complex three-dimensional features, such as salt diapirs, this can be incredibly important. The geoscientist should be aware of the overburden within the seismic data. Another possible clue that anisotropy needs to be captured is if well ties consistently mis-tie the target interval or reservoir depth predictions are consistently wrong.

To understand anisotropy in an area of interest, a good starting place is with well-based velocities which can give a first pass indication of the variability compared to the isotropic model. Thomsen's parameters can be calculated in the wells to give ε and δ values which will indicate the percentage of anisotropy present in specific intervals. Starting migration testing with a comparison of isotropic against VTI or TTI velocity model allows the processor and geoscientist to observe whether the image evolution is moving in the right direction and, consequently, make further adjustments before moving to the more complex orthorhombic velocity model. Where a data set has been migrated with isotropic velocities it is always possible to update the velocity model using a combination of ε and δ from well data and use tomography methods to update the velocity model before re-migration.

AVO

Anisotropy can affect AVO analysis in several ways. The analysis of AVO relies on observational variations in seismic amplitude based on variable incident ray angles away from vertical at a reflection boundary. It is, therefore, likely that seismic amplitudes used for AVO analysis will be sensitive to anisotropy effects (Blangy 1994). Standard AVO equations and approximations assume P-wave reflections from isotropic media above and below the interface boundary. Rüger (1997) discusses the impact of anisotropy in AVO analysis and proposes an amended version of the linearized approximation of the Zoeppritz equation by Aki-Richards as a solution. The solution is valid for weak anisotropy and both VTI and HTI media. Ruger (1998) takes this further to include the effects of orthorhombic anisotropy and shear-wave splitting on amplitude and AVO gradient calculations.

Seismic reflection event flatness observed in offset gathers is directly related to the correction applied to the velocity. Isotropic velocities are unable to sufficiently flatten gathers if they require an anisotropic velocity field, which can then lead to inaccurate amplitude measurements with offset of attributes such as AVO gradient. When looking at AVO in clastic systems it is common for a large, thick shale overlying the hydrocarbon reservoir sand to act as a sealing lithology. The sealing shale at a hydrocarbon accumulation must be of sufficient thickness to hold back the pressures of the estimated hydrocarbon column and make a spatially continuous lateral seal. As discussed, shales are inherently anisotropic due to their mineralogical composition. One-dimensional half space modelling along with well-based synthetic AVO modelling are commonly used methods for assessing the presence and viability of an AVO signal. If the overlying shale is anisotropic then the predicted AVO curves from the half-space and the synthetic gathers will not correctly match the real seismic observations. This means that validating an observed AVO signal in the data becomes difficult and introduces greater uncertainty. Asaka (2018) shows a real data example from a case study using wells from the Browse Basin, Western Australia, along with evidence of the effects of anisotropy on AVO gradient.

Seismic AVO inversion is also reliant on variations of seismic amplitudes and gradient calculations, and, generally, workflows rely on anisotropy to be dealt with during seismic data processing. When considering anisotropy, small angle ranges showing VTI are unlikely to show variability. However, when larger angle ranges are used as input into the inversion the anisotropy

will likely become an issue that must be addressed (Sams and Annushia 2018).

4.7 Attenuation—Q Factor

Attenuation is the reduction of seismic wavefield energy as it propagates through the subsurface. Attenuation effects the high-frequencies more that the low-frequencies because their shorter wavelengths are more abundant per cycle compared to longer waves. As high-frequencies are reduced in the signal the dominant signal wavelength and period become longer which results in a reduction of resolution within the data. Attenuation also has an effect of distorting the phase of seismic wavelet phase within the data.

Wave attenuation is described using an anelastic attenuation factor, commonly referred to as the seismic quality factor, Q factor or simply Q. Q is a dimensionless factor linked to the physical state of the rock and is defined by the ratio of the maximum energy stored during a cycle divided by the energy lost during a cycle (Kjartansson 1979).

There are several mechanisms that cause attenuation. Reduction of transmitted energy occurs at interfaces in the subsurface where reflections, diffractions, scattering and mode conversions are generated. The larger the reflection coefficient the greater the attenuation of energy for transmitted waves. Spherical divergence of the wavefront moves largely away from the source, which causes a spherical spreading and dispersive attenuation of the wave energy away from the source point. Propagation of acoustic waves through the subsurface (Chap. 2) involves transmission through particles in the subsurface. This vibration of particles leads to the generation of frictional heat which is a mechanism for attenuation of energy. The level of energy lost in this way is highly dependent on the type of lithology and the associated porosity and fluid properties of the subsurface rocks the waves are travelling through. Taking all these different mechanisms together the general

Table 4.1 Showing Intrinsic attenuation measurements in rock types (Yilmaz 2001)

Rock type	Attenuation constant, Q	Frequency range, Hz
Basalt	550	3,000–4,000
Granite	300	20,000–200,000
Limestone	50–650	2–18,000
Chalk	2	150
Shale	15–75	3,300–75,500
Sandstone	25–125	2–30,000

statement can be made that longer travel time of acoustic waves in the subsurface leads to increased attenuation:

$$Q = 2\pi / \left(\frac{\text{cycle max energy}}{\text{cycle energy lost}} \right). \qquad (4.7)$$

When carrying out any type of quantitative analysis using seismic amplitudes such as AVO (Sect. 6.1), the effect of attenuation can be an issue for the interpreter in understanding the true seismic signal. If a robust and accurate estimation of Q factor can be obtained, then a correction can be made to the data. The negative effects of attenuation for AVO analysis is also being researched to support de-risking DHIs, especially when dealing with low saturation gas which can fool conventional AVO techniques (Chapman and Liu 2006).

The method of applying a correction for Q requires a measurement that can be obtained from VSP data, well-logs and laboratory measurements using core or seismic data. Q values vary greatly for specific rock types based on laboratory measurements (Table 4.1), so although the laboratory measurements may a useful reference, the application of a Q correction to seismic data relies on other data sources. It can be challenging to have sufficient VSP or well data to build a heterogeneous Q model. Methods based on post-stack data are usually produced in the frequency domain which leads to averaging effects (Sun et al. 2014). As such, an estimation of Q in pre-stack data can be common, particularly when the correction is applied prior to or during migration.

Inverse Q filtering is the application of an inversion operator that counteracts the absorption and dispersion effects. This process requires estimating Q from the seismic data in shallow areas where attenuation is likely to be lower and, so, contain more high-frequencies when compared with deeper seismic traces where more high-frequencies have been attenuated. The smaller the Q value obtained from the data the greater the time dependant decrease in high-frequencies relative to its low-frequencies (Hill and Rüger 2020). The presence of noise and multiple energy in the data can cause inaccurate estimates of Q, so data must undergo denoise processes prior to applying a Q estimation and application workflow. The effect of applying the inverse Q can be seen in the frequency spectra differences before and after application (Fig. 4.19).

The interpreter should be aware that the ability to retrieve bandwidth caused by Q attenuation using inverse Q filtering is limited by frequency-dependent signal-to-noise ratio. The Eq. 4.8 estimates the highest retrievable wavelet frequency (Hill and Rüger 2020). This means that the recovery of frequencies is still bound to the fact that larger travel times will affect the ability to retrieve higher frequencies within the data:

$$F_{Max} = D.\frac{Q}{t}, \qquad (4.8)$$

where F_{Max} is the maximum frequency of the signal, D is the target depth, Q is the Q factor and t the TWT at the target.

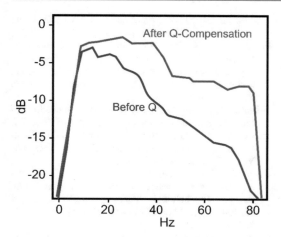

Fig. 4.19 The effect on frequency of applying an inverse Q filter

The application of inverse Q occurs for several reasons, for example, when the de-multiple processes have been applied too aggressively which damages the primaries and makes the seismic data look low amplitude and washed out. Another instance is where imaging is poor due to illumination issues around salt diapirs and large faults. The last situation is where a well-based synthetic model shows an expected AVO response which is not observed in the seismic data. Altering the seismic response to match well-based synthetics is always hazardous and if not applied carefully runs the very real risk of introducing erroneous features in the data. In acknowledgement of this pitfalls several different amplitude friendly workflows for applying Q application can be found in various publications: Trantham and He (2017) and Seher et al. (2019).

Although inverse Q filtering can be applied to pre-stack data using a time varying method to compensate for variation in the level of attenuation, the filter is unable to deal with differing amount of absorption along complex ray paths in a complex medium. The inverse Q does not account for complex structures and dipping reflectors, and it struggles to apply the attenuation compensation fully in three dimensions (Jinliang et al. 2016). To overcome this issue Q compensation techniques can be incorporated as an additional feature to several other important seismic processing flows, such as velocity tomography and migration. For migration the Q compensation is applied along migrated ray paths to boost the high-frequencies, which results in a better image with more coherent events and increased resolution due to the frequency recovery. Migration Q can be applied in both vertical and horizontal directions because ray paths are found at a variety of angles. Applying the Q compensation within pre-stack depth migration gives a more geological appropriate application compared to the application of inverse Q.

The interpreter should carefully assess the strength of any Q attenuation workflow applied to the data using stacked volumes at different offsets or angles. Amplitude extractions at key seismic horizons are also useful for the QC of Q attenuation processes. Q attenuation workflows done before migration will be costly and time consuming to change once a final processed volume has been generated, therefore the interpreter should feel comfortable with what has been applied before a processing workflow moves on. Post-migration Q workflows are easier to reverse. However, it is advisable to retain gathers or stacks with and without any post-migration amplitude or frequency applications.

References

Abma R, Claerbout J (1995) Lateral prediction for noise attenuation by t-x and f-x techniques. Geophysics 60:1887–1896

Abma R, Kabir N, Matson KH, Michell S, Shaw SA, McLain B (2005) Comparisons of adaptive subtraction methods for multiple attenuation. Lead Edge 24:277–280

Asaka M (2018) Anisotropic AVO: implications for reservoir characterization. Lead Edge 37:916–923

Baden D, Henry P, Saracco G, Marié L, Tonetto A, Guglielmi Y, Nakagawa S, Massonnat G, Rolando J-P (2017) How seismic anisotropy changes with scale. SEG technical program expanded abstracts, pp 305–309

Berkhout AJ, Verschuur DJ (2003) Transformation of multiples into primary reflections. SEG technical program expanded abstracts, pp 1925–1928.

Berkhout AJ, Verschuur DJ (2005) Removal of internal multiples with the common-focus-point (CFP) approach: part 1—explanation of the theory. Geophysics 70(3)

Blangy JP (1994) AVO in transversely isotropic media—an overview. Geophysics 59:775–781

Brittan J, Wrench A (2004) Attenuation of multiple diffractions using a cascaded noise removal sequence. SEG Int'l exposition and 74th annual meeting Denver Colorado 10–15th Oct 2004

Brookes D (2011) Case studies in 3D interbed multiple attenuation. Lead Edge 30:914–918

Brown MP, Guitton A (2005) Least-squares joint imaging of multiples and primaries. Geophysics 70:S79–S89

Canales L (1984) Random noise reduction. 54th annual international meeting, SEG, expanded abstracts, pp 525–527

Casasanta L, Telling R, Grion S (2020) De-signature of apparition-blended seismic data: a North Sea example. In: SEG international exposition and 90th annual meeting. https://doi.org/10.1190/segam2020-3427953.1

Chapman M, Liu E (2006) Seismic attenuation in rocks saturated with multi-phase fluids. In: SEG technical program expanded abstracts, pp 1988–1992

Chen K, Sacchi MD (2013) Robust reduced-rank seismic denoising. In: SEG Houston 2013 annual meeting

Davison CM, Poole G (2015) Far-field source signature reconstruction using direct arrival data. In: 7th EAGE conference & exhibition 2015. IFEMA Madrid, Spain 1–4 June 2015

de Jonge T, Vinje V, Poole G, Hou S, Iversen E (2022) Debubbling seismic data using a generalized neural network. Geophysics 87:V1–V14

Deng J, Wang C, Zhao Q, Guo W, Tang G, Zhao J (2021) Depositional and diagenetic controls on macroscopic acoustic and geomechanical behaviors in Wufeng-Longmaxi formation shale. Front Earth Sci

Dix CH (1955) Seismic velocities from surface measurements. Geophysics 20:68–86

Douma H, Jenner E, Kumar R, Al-Kanderi J (2014) Rayleigh-wave filtering through phase-velocity dispersion inversion and modeling: application to north Kuwait 3D seismic field data. In: 2014 SEG Denver annual meeting

Dragoset B (2000) Introduction to air guns and air-gun arrays. The Leading Edge 19(8): 817–928

Dutta G, Huang H, Kanakamedala K, Deng B, Wang P (2019) Practical strategies for interbed multiple attenuation. In: SEG international exposition and 89th annual meeting

Elboth T, Qaisraniand HH, Hertweck T (2008) Denoising seismic data in the time-frequency domain. In: SEG Las Vegas 2008 annual meeting

Ernst FE, Hermanz GC, Ditzel A (2002) Removal of scattered guided waves from seismic data. Geophysics 67(4):1240–1248

Etris EL, Crabtree NJ, Dewar J (2002) True depth conversion: more than a pretty picture. CSEG Recorder 26(09)

Fomel S (2006) Towards the seislet transform. In: SEG New Orleans 2006 annual meeting. https://doi.org/10.1190/1.2370116

Francis A (2018) A simple guide to seismic depth conversion: part 1. GeoExpro 15(2)

Fu Q, Luo Y, Kelamis PG, Huo S, Sindi G, Hsu S-Y, Weglein AB (2010) The inverse scattering series approach towards the elimination of land internal multiples. In: SEG technical program expanded abstracts, pp 3456–3461

Griffiths M, Hembd J, Prigent H (2011) Applications of interbed multiple attenuation. Lead Edge 30:906–912

van Groenestijn G-J, Ross W, Cumaran GN (2012) Connecting deconvolution and SRME. SEG Las Vegas 2012 annual meeting

Guan H, Li G, Wang B, et al.(2008) A multi-step approach for efficient reverse-time migration78th Annual International Meeting, SEG, Expanded Abstracts, pp 2341–2345, https://doi.org/10.1190/1.3059350

Guan H, Dussaud E, Denel B, Williamson P (2011) Techniques for an efficient implementation of RTM in TTI media. In: SEG Technical program expanded abstracts, pp 3393–3397

Gulunay N (1986) F-X decon and the complex Weiner prediction filter for random noise reduction on stacked data. In: Society of exploration geophysicists 56th annual international meeting, Houston, TX

Hampson D (1986) Inverse velocity stacking for multiple elimination. J Canadian Soc Explor Geophys 22(1):44–55

Hargreaves N, Wombell R (2004) Multiple diffractions and coherent noise in marine seismic data. In: SEG Int'l exposition and 74th annual meeting Denver Colorado 10–15th Oct 2004

Hill SJ, Rüger A (2020) Illustrated seismic processing volume 1&2: imaging. Society of Exploration Geophysicists, Tulsa, OK

Hlebnikov V, Elboth T, Vinje V, Gelius L-J (2021) Noise types and their attenuation in towed marine seismic: a tutorial. Geophysics 86(2):W1–W19

Holt RA, Campbell D (2008) Estimating lateral positioning uncertainty after anisotropic depth migration: a thrust belt case history. In: SEG technical program expanded abstracts, pp 232–236

Ikelle LT (2005) A construct of internal multiples from surface data only: the concept of virtual seismic events. Geophys J Int 164:383–393. https://doi.org/10.1111/j.1365-246X.2006.02857.x

Jakubowicz H (1998) Wave equation prediction and removal of interbed multiples. In: SEG technical program expanded abstracts, pp 1527–1530

Jinliang X, Wenliang Z, Bing L, Guowang Z, Hongguo Q, Shuyuan D (2016) Application of Q-PSDM on land data, a case study from East China. In: SEG global meeting abstracts, pp 589–591

Jones IF (2018) Velocities, imaging, and waveform inversion—the evolution of characterizing the Earth's subsurface. In: 2018 EAGE education tour series

Kendall J-M, Fisher QJ, Crump SC, Maddock J, Carter A, Hall SA, Wookey J, Valcke SLA, Casey M, Lloyd G, Ismail WB (2007) Seismic anisotropy as an indicator of reservoir quality in siliciclastic rocks, vol 292.

Geological Society, London, Special Publications, pp 123–136

Kjartansson K (1979) Constant Q-wave propagation and attenuation rock physics project. J Geophys Res 84 (B9):4737–4748

Landrø M (1992) Modelling of GI gun signatures. Geophys Prospect 40:721–747

Landrø M, Amundsen L (2010) GeoExPro. Marine Seismic Sources Part I 7(1)

Liu Y, Fomel S (2010) OC-seislet: seislet transform construction with differential offset continuation. Geophysics 75: WB235–WB245

Liu E, Martinez A (2012) Seismic fracture characterisation concepts and Practical applications. Education tour Series EAGE

Liu Y, Chang X, Jin D, He R, Sun H, Zheng Y (2011) Reverse time migration of multiples for subsalt imaging. Geophysics 76: WB209–WB216

Liu Y, He B, Zheng Y (2020) Full-waveform inversion using multiples and primaries. In: SEG technical program expanded abstracts, pp 3838–3842

Lokshtanov D (1999) Multiple suppression by data-consistent deconvolution. Lead Edge 18(1):115–119

Luo Y, Kelamis PG, Fu Q, Huo S, Sindi G, Hsu S-Y, Weglein AB (2011) Elimination of land internal multiples based on the inverse scattering series. Lead Edge 30:884–889

Ma C, Guo M, Liu Z, Sheng J (2020) Analysis and application of data-driven approaches for internal-multiple elimination. In: SEG technical program expanded abstracts, pp 3124–3128

Matson K, Corrigan D, Weglein A, Young C-Y, Carvalho P (1999) Inverse scattering internal multiple attenuation: Results from complex synthetic and field data examples. In: SEG technical program expanded abstracts, pp 1060–1063

Mojesky T, Loh TC, Eliott-Lockhart R (2013) Modelling and removing WAZ OBC interbed multiples. In: ASEG extended abstracts, vol 1, pp 1–4. https://doi.org/10.1071/ASEG2013ab103

Muijs R, Robertsson JO, Holliger K (2007) Prestack depth migration of primary and surface-related multiple reflections: part I—imaging. Geophysics 72:S59–S69

Naghizadeh M, Sacchi M (2018) Ground-roll attenuation using curvelet downscaling. Geophysics 83(3):V185–V195

Nekut AG, Verschuur DJ (1998) Minimum energy adaptive subtraction in surface-related multiple attenuation. In: Delft Univ. of technology 1998 SEG expanded abstracts

Osen A, Amundsen L, Reitan A (1999) Removal of water-layer multiples from multicomponent sea-bottom data. Geophysics 64(3):838–851

Osen A, Amundsen L (2001) Multidimensional multiple attenuation of OBS data. In: SEG international exposition and annual meeting San Antonio, Texas, 9–14 Sept 2001

Pham N, Li W (2022) Physics-constrained deep learning for ground roll attenuation. Geophysics 87(1):V15–V27

Pica A, Delmas L (2008) Wave equation based internal multiple in 3D. In: SEG technical program expanded abstracts, pp 2476–2480

Ramírez AC, Weglein AB (2008) Inverse scattering internal multiple elimination: leading-order and higher-order closed forms. In: SEG technical program expanded abstracts, pp 2471–2475

Ras P, Volterrani S, Walz A, Hannan A, Narhari SR, Al-Ashwak S, Kidambi VK, Al-Qadeeri B (2012) Interbed demultiple workflow for Kuwait's deep carbonate reservoirs. In: SEG global meeting abstracts, pp 1–4

Rüger A (1997) P-wave reflection coefficients for transversely isotropic models with vertical and horizontal axis of symmetry. Geophysics 62:713–722

Rüger A (1998) Variation of P-wave reflectivity with offset and azimuth in anisotropic media. Geophysics 63: 935–947

Sacchi MD, Ulrych TJ (1995) High-resolution velocity gathers and offset space reconstruction. Geophysics 60 (4):1169–1177

Sams M, Annushia A (2018) Inversion in a VTI medium. In: SEG technical program expanded abstracts, pp 436–440

Scholtz P, Masoomzadeh H, Camp R (2015) Directional designature without near-field hydrophone recordings. In: SEG New Orleans annual meeting. https://doi.org/10.1190/segam2015-5843268.1

Seher T, Kokoshina E, Spoors S (2019) Amplitude friendly inverse Q filtering. In: SEG technical program expanded abstracts, pp 625–629

Sheriff RE (1991) Encyclopedic dictionary of exploration geophysics, 3rd edn. Society of Exploration Geophysicists, Tulsa, USA

Singh Y (2012) Deterministic inversion of seismic data in the depth domain. Lead Edge 31:538–545

Souza AJ, Loures LG (2009) The Curvelet Transform for ground-roll suppression. In: 11th international congress of the Brazilian geophysical society, Brazil, 24–28 Aug 2009

Spitz S (1999) Pattern recognition, spatial predictability, and subtraction of multiple events. The Leading Edge 18: 55–58

Staring M, Wapenaar K (2020) Three-dimensional Marchenko internal multiple attenuation on narrow azimuth streamer data of the Santos Basin. Brazil Geophys Prospect 68:1864–1877

Strobbia C, Vermeer P, Laake A, Glushchenko A, Re S (2010) Surface waves: processing, inversion and removal. First Break 28(8)

Sun SZ, Wang Y, Sun X, Yue H, Yang W, Li C (2014) Estimation of Q-factor based on prestack CMP gathers and its application to compensate attenuation effects. In: SEG technical program expanded abstracts, pp 3709–3714

Thomsen L (1986) Weak elastic anisotropy. Geophysics 51(10): 1954–1966

Thorbecke J, Zhang L, Wapenaar K, Slob E (2021) Implementation of the Marchenko multiple elimination algorithm. Geophysics 86:F9–F23

Torabi S, Javaherian A (2012) Ground roll attenuation by focal transform. Istanbul International Geophysical Conference and Oil & Gas Exhibition, Istanbul, Turkey, 17–19 Sept 2012.

Trantham EC, He K (2017) AVA friendly Q amplitude compensation. In: SEG international exposition and 87th annual meeting. https://doi.org/10.1190/segam2017-17073710.1

Trickett S (2008) F-xy Cadzow noise suppression. In: SEG Las Vegas 2008 annual meeting

Van Borselen RG, Schonewille MA, Hegge RF (2005) 3D surface-related multiple elimination: acquisition and processing solutions. Lead Edge 24(3):260–268

Verschuur DJ, Berkhout AJ, Wapenaar CPA (1992) Adaptive surface-related multiple elimination. Geophysics 57(9):1166–1177

Verschuur D, Berkhout A (2005) Removal of internal multiples with the common-focus-point (CFP) approach—part 2: application strategies and data examples. Geophysics 70(3):V61–V72

Weglein AB, Boyse WE, Anderson JE (1981) Obtaining three-dimensional velocity information directly from reflection seismic data: an inverse scattering formalism. Geophysics 46(8): 1116

Weglein AB, Nita BG, Innanen KA, Otnes E, Shaw SA, Liu F, Zhang H, Ramírez AC, Zhang J, Pavlis GL, Fan C (2006) Using the inverse scattering series to predict the wavefield at depth and the transmitted wavefield without an assumption about the phase of the measured reflection data or back propagation in the overburden. Geophysics 71(4): SI125

Weglein AB (2009) A new, clear and meaningful definition of linear inversion: implications for seismic inversion of primaries and removing multiples. In: SEG technical program expanded abstracts, pp 3059–3063

Winterstein DF (1986) Anisotropy effects in P-wave and SH-wave stacking velocities contain information on lithology. Geophysics 51:661–672

Wood LC, Heiser RC, Treitel S, Riley PL (1978) The debubbling of marine source signatures. Geophysics 43(4):715–729

Yang Z, Hembd J, Chen H, Yang J (2015) Reverse time migration of multiples: applications and challenges. The Leading Edge 34: 780–782, 784–784, 786–786

Yarham C, Boeniger U, Herrmann F (2006) Curvelet-based ground roll removal. In: SEG New Orleans 2006 annual meeting

Yilmaz O (2001) Seismic data analysis. Society of Exploration Geophysicists, Tulsa, OK, USA

Yuan Y, Si X, Zheng Y (2020) Ground-roll attenuation using generative adversarial networks. Geophysics 85 (4): WA255–WA267

Ziolkowski A, Parkes G, Hatton L, Haugland T (2983) The signature of an air gun array: computation from near-field measurements including interactions. Geophysics 47:1413–1421

Seismic Well Ties and Wavelets

Wavelets play a critical role in understanding how to relate seismic reflections to geology in the subsurface. A wavelet can be thought of as a discrete packet of energy with a defined length, amplitude, phase and frequency content, and it is a numerical representation of the acoustic energy generated at the source location.

A wavelet acts as a filter allowing the geoscientist to view the geology encountered by the seismic energy from the source as it travelled through the subsurface. Wavelets are used in a range of geophysics and geo-modelling workflows dealing with seismic data, but before being used they must be extracted, or estimated, from the recorded geophysical data. The impact of how this extraction and estimation is done and the resulting wavelet obtained is often overlooked as being trivial, which is far from true. Subtleties in the wavelet characteristics can significantly alter both a seismic image and the results of the quantitative and qualitative seismic interpretation and should therefore be a point of focus for the geoscientist.

5.1 Seismic-To-Well Tie

The process of seismic-to-well tie is one of the most fundamental and important processes carried out by a geoscientist for interpretation, seismic forward modelling or seismic inversion. Done properly the process sets up a project to be accurate and increases the understanding of the subsurface along with subtle and important geological and depositional features.

Although there are number of partially and fully automated well-tie algorithms available (Herrera et al., 2014; Nivlet et al., 2020), it is widely accepted that human intervention is a required part of the well-tie process. The reason being that tying wells goes beyond the analytic autocorrelation performed automatically by the software application; an understanding about the characteristics of the background geology and the seismic data must be included. Figure 5.1 shows a schematic representation of the well-tying process. The geoscientist can choose to display a range of available well-logs during the well tie that might aid in the visual alignment of the well synthetic trace and seismic trace. Fig. 5.1 shows an acoustic impedance log, which is calculated by using the displayed P-velocity log and density log (Eq. 2.15). The reflection coefficient log (RC) represents the P-wave reflection coefficient at a normal incidence angle (Eq. 2.16). Finally the convolution process shown using the estimated wavelet, which produces the synthetic trace, is performed in the time domain.

Wavelets are particularly important for performing the task of seismic-to-well ties (e.g., generating well-log based synthetic seismic traces that can be compared to processed seismic traces around the well). The process of tying a well to seismic data is required because the well data is a direct measurement of the geology in the

© The Author(s), under exclusive license to Springer Nature Switzerland AG 2022
T. Tylor-Jones and L. Azevedo, *A Practical Guide to Seismic Reservoir Characterization*,
Advances in Oil and Gas Exploration & Production, https://doi.org/10.1007/978-3-030-99854-7_5

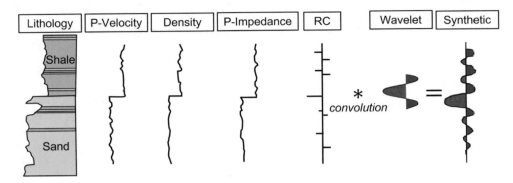

Fig. 5.1 Schematic representation of the well-tying process, showing the idea behind using the velocity and density logs to produce reflections that allow synthetic seismic data to be generated

subsurface in depth. These well data give accurate, descriptive and analytical information about the types of geology and fluids encountered in the borehole. These data then be used to date subsurface intervals and describe depositional environments at depths of interest. By linking this well-based interpretation to the peaks and troughs in the reflectivity seismic data we can extrapolate the interpretation over the spatially much larger seismic data.

Because we have access to the exact position of the well head, and a good estimation of the well deviation path, depths are accurately measured along the wellbore and the acquired

well-logs have high vertical resolution (1–3 ft). Unfortunately, the spatial density of well data can often be sparse in an area of interest, especially offshore. Onshore fields sometimes have greater well density from the practice of pattern drilling for field development. On the other hand, seismic reflection data covering large areas are indirect measurements of the subsurface and have much lower vertical resolution with uncertainties about the positioning of reflectivity events (Fig. 5.2).

To take the extensive and detailed knowledge of the well along the vertical direction and match it to events in the seismic data comes with one

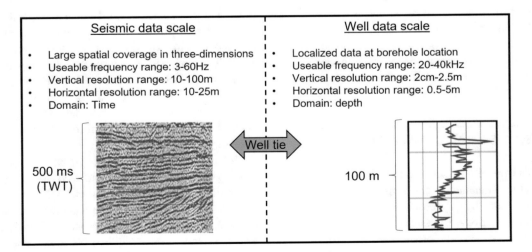

Fig. 5.2 Comparison between the characteristics of the seismic and well data

main issue: the well is in the depth domain and the seismic is, (unless deep migrated) in the time domain. To overcome this limitation, a time-to-depth relationship must be generated with which to accurately compare the two data types in the same domain.

Convolution is a mathematical operation commonly used in seismic processing and interpretation. A signal (reflection coefficient) is used as input data and is convolved with an operator (i.e., the wavelet) which acts like a filter to change the shape of the input signal to produce an output signal (i.e., seismic trace). The actual convolution process can be thought of as the summation of a series of multiplications. Figure 5.3 shows a comparison between the convolution model used to generate a seismic trace and the process of wavelet estimation for well tying. In the convolution model, the seismic trace is a product of the reflection coefficient of the earth convolved with a wavelet, which is defined by the input pulse of the source. For the wavelet estimation, we take the reflection coefficient calculations from the well acoustic impedance logs derived from velocity (sonic log) and density. The reflection coefficient log is then convolved with an estimated wavelet to produce a synthetic trace. With the wavelet estimation method, we must iteratively work out the most accurate wavelet using the match between the synthetic trace from the well and the real seismic data as a check.

Before estimating a wavelet and performing a well tie data quality control analysis must be carried out on the available well data. This phase can be very time-consuming compared to the actual task of tying the wells, depending on the number of wells available. However, it is a critical part of the workflow because it directly influences the quality of the well tie and provides important insights into previous drilling activity and geological understanding in the area of interest, while adding to the geoscientist's broader knowledge of the subsurface.

5.1.1   Well Data Gathering and Quality Control

The first stage of the well-tie process is to identify the well data that are available within the area of interest. To carry out AVO forward modelling (Chap. 6) or seismic inversion (Chap. 7), the wells ideally need to have recorded sonic (acoustic and shear) and density logs within the survey area. Generally the wells are tied to a three-dimensional seismic survey; however, it is also possible for the area of interest to include two-dimensional seismic lines. Where only two-dimensional data are available it is important to find out whether the well(s) you are looking to tie falls exactly on the two-dimensional seismic lines. If not, which is usually the case, a

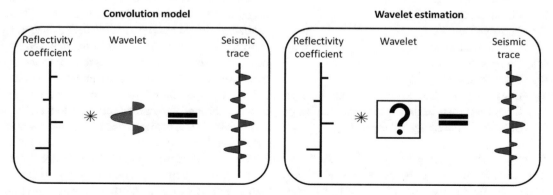

Fig. 5.3 Relationship between the convolution model and the wavelet estimation method

measurement must be made to assess how far the data are apart: 10, 100, 500 m? Sometimes well data are outside the coverage of the available three-dimensional seismic survey, but if an important geological interval has been penetrated or it is close enough to a key area the data may be considered useful. Additionally, geology that appears horizontally flat is likely to extend beyond the edge of the seismic data footprint to the well. In a frontier exploration setting with very little data, you could justify tying a well from outside the seismic survey extents providing it is highlighted as an interpretation risk/uncertainty. A judgement call from the geoscientist will be required as whether or not to artificially move the well data and tie to the nearest seismic traces. This is usually only a consideration for a particularly key well or when well data are very sparse.

Once the initial well data set has been decided upon the next step is a log data summary of each well. This step is critical as it defines how much additional work and/or expertise is needed before each well is suitable for tying to the seismic data. This stage can also refine the well selection, which initially used the criteria of well proximity to the area of interest. However, on closer inspection the data quality may be insufficient to proceed with. A common, but not exhaustive, list of required data might be:

(a) List of logs (LWD and wireline);
(b) Checkshot or vertical seismic profile (VSP);
(c) Well reports, sampling report, special analysis reports (e.g., SCAL);
(d) Deviation surveys;
(e) Tops/well markers;
(f) CPI of well data;
(g) Composition logs;
(h) Petrophysical logs (S_w, $\phi.V_{sh}$);
(i) Logs that indicate borehole conditions as these may help to understand the quality of other log curves (e.g., Caliper log).

In cases where the existing wells have different sets of well-logs or well-logs were not acquired for the section of interest, well-log imputation techniques might be used to predict the missing information. These techniques might be as simple as linear regressions to more advanced methods based on data science and machine learning. An important aspect to consider is the prediction of the well'log plus an uncertainty range (Feng et al. 2021).

Well markers are an interpretation based on recorded well-log data and indicate the top, base or intermediate divisions of a particular geological unit or sequence (Fig. 5.4). They provide the geoscientist with information about lithology changes in a well-log and the geological age of the lithology. Well markers are most commonly interpreted using the well data response of multiple recorded logs or biostratigraphy samples obtained from the cuttings or cores from the well. Ideally a combination of both techniques helps to give a more integrated interpretation. An established and understood regional geological framework underpins the interpretation of well markers in all but the most frontier exploration areas. Depending on the technical requirements of a workflow the number of stratigraphic subdivisions using well markers varies. Displaying the right level of detail is important so data can be clearly understood and visualized.

When comparing well markers interpretated from just log responses to those from just biostratigraphy analysis there will likely be differences in their depth position within the well, from a few metres to tens of metres. It should be recognized that when using well markers in well ties some geological markers may not match up with the seismic and synthetic events, which are dependent on relatively strong contrasts in elastic properties to produce reflectivity events. Consideration should also be given to the consistency of well markers used across all wells; they should be spaced vertically at an appropriate distance to match the resolution of the seismic data, which helps the alignment of discrete peak and trough events during interpretation. Geophysicists, geologists and petrophysicists should work together on appropriately defining well markers. Within a joint-venture project the consistent use of the same well markers and nomenclature, rather than sticking to internal naming conventions, will help avoid confusion.

The final stage of well data preparation is a thorough petrophysical analysis carried out on all

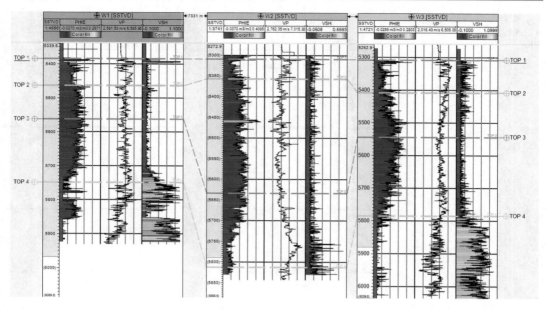

Fig. 5.4 Example of three wells with markers defining the top of lithofluid facies. The left track exhibits effective porosity, the middle track V_P and the right track volume of shale. Vertical axis in metres

wells prior to use in a well tie. Although well ties can be performed without this step it is highly recommended to assure validity of the final results. Petrophysical analysis will help to decrease, or better assess, uncertainties related to the data and lead to a detailed understanding of not only what a well has encountered but how the drilling operation of the well went, which can be very useful information. The analysis provides information about the presence of hydrocarbon shows, areas of faulting and fractures, identifying poorly consolidate formations, formation water composition and, very importantly, pressures and temperatures. Additionally, the petrophysics will be able to repair and correct issues with the recorded logs and ensure consistency from each well. The standard outputs from a petrophysical analysis are calculated logs for clay fraction, water saturation and porosity, both total and effective.

order that the well and seismic data can be displayed in the same vertical domain. The well-log is brought to the time domain simply because there is more certainty in the measurement of the well-logs in depth. To do the inverse (i.e., bring the seismic data to the depth domain) is a longer, slower process and with more uncertainties. To create the initial time-depth relationship, data from the well-logs is required along with a checkshot survey or VSP (Fig. 5.5). However, if unavailable a P-sonic log or velocities from seismic processing can be used as a starting time-depth relationship.

Carrying out a VSP should always be included as contingency in the cost of a well program due to their valuable source of additional information, beyond just use in well ties. Following other processes in the well-tie workflow, quality control of the checkshot or VSP data is required along with a level of editing.

5.1.2 Initial Time-Depth Calibration

To begin the well-tying process the geoscientist needs to create a time-to-depth relationship in

5.1.3 Well-Log Upscaling

As mentioned earlier in the chapter, there is a much higher vertical resolution in the well data

Fig. 5.5 Difference between:
a checkshot survey and **b**
VSP

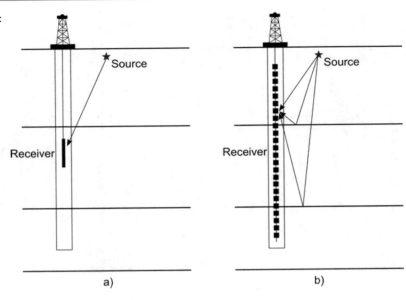

a)   b)

compared to that of the recorded seismic data. For the seismic well tie, too much variability in the well-logs of sufficient magnitude results in the generation of lots of reflectivity events during the convolution with the estimated wavelet. This can produce synthetic traces that appear noisy and slightly chaotic in appearance, so the geoscientist should consider whether to upscale well-logs to something closer to the vertical seismic scale. This option is often used when the well-logs (e.g., velocity and density) are particularly variable caused by multiple thin beds with highly contrasting acoustic impedance. An example would be from a tertiary interval of the UK North Sea where dense cluster of cemented (<1m individual thickness) are often present in sandstone reservoir formations, causing the well-log response to appear spiky. A despike tool can remove these events but the log may still contain too much fine-scale complexity to achieve a good well tie. The remedy is to upscale the well-logs to remove some of the complexity and noise and leave the larger scale boundaries within the log, which will match those seen by the seismic data. A common upscaling methodology is Backus averaging (Lindsay and Koughnet 2001) which allows a vertical thickness window length to be specified. This sliding window is applied to produce a set of filtered logs or it can be applied

within a well-tie workflow before the reflection coefficients are calculated. Along with upscaling other smoothing and filtering operators can be applied to condition the logs to be more suitable for the well tie. The seismic wavelet used in the convolution process itself can be used as a downscaling filter.

There are instances where the finer detail from the well-log is of critical value to the well tie and the seismic is actually not seeing them. In this case, the well-log can be used to support spectrally enhancing the seismic to provide a better match to the events in the well synthetic traces (Jilinski, 2015).

5.1.4   Selection of the Seismic

The next data decision to consider in the well-tie workflow is the seismic data. The volume selected for the well tie will be compared to the synthetic traces coming from the well reflectivity convolved with the estimated wavelet, and it will be the primary volume used for seismic interpretation. Some simple considerations will aid in the selection of the most suitable volume. The seismic data should spatially cover all the wells to be tied and include data down to and, if possible, slightly below the interval of interest to ensure an optimum well tie. Production effects caused by development activities can alter the character of seismic data, in this case the data used for well ties should be of a suitably modern vintage. Basic knowledge of the acquisition and processing parameters should be known about the volumes being selected for the well tie. There might be a few different volumes to choose from with different acquisitions (e.g., multi-client surveys and proprietary surveys) and processing workflows applied.

The newest survey or most recently re-processed seismic data are not always the best for using in the well tie. As mentioned in (Chap. 3). Newer seismic technologies, such as OBN and OBC, are more complex to process and can require newer workflows that are not routinely used in standard conventional streamer data processing. Broadband data can yield very different looking final results and if not processed properly can appear of lower quality than older data. It is always good practice to make a technical summary of the different seismic volumes so that the decision is informed, and all the information can be compared in one document. The main categories that will determine the final selection fall into four broad categories: acquisition parameters, processing parameters, visual appearance and analytical investigation. Although broad, these categories provide a framework from which to investigate the available seismic (Table 5.1).

The suggestions in Table 5.1 are not intended to challenge the quality control steps coming from an acquisition or processing project. They are simply areas of information and investigation that will give the geoscientist a better understanding of the seismic data and make the choice of data more analytically informed. It is likely that two volumes will have a range of pros and cons that make them both contenders. In this case, performing a well tie and wavelet estimation with both volumes for a few wells will help identify the most suitable one, bearing in mind the objectives of the interpretation. Another option that may also enable the geoscientist to identify the preferred volume is to look at how they respond to some simple post-stack enhancement processes. Processes such as structural filtering, spectral shaping and bandlimited relative inversion (e.g., coloured inversion) can bring great additional value to a data set beyond its current state, and two data sets will often behave differently due to their properties and characteristics. The best volume selection after these processes will sometimes not be what is expected so it is important to stay open minded and unbiased throughout the analysis, alongside sharing findings and seeking specialist opinions before making a final decision.

If we are tying seismic data to obtain a wavelet for seismic inversion further thought is required around the seismic data to be used for well ties. A partial-stack inversion that produces S-impedance and V_P/V_S ratio may require multiple wavelet estimations for each partial stack to correctly account for variations in frequency content. Some inversion workflows avoid this by carrying out a basic frequency-balancing workflow across all the partial-stacks to a chosen reference stack. However, if individual partial-stack wavelets are required the first consideration is the offset or

Table 5.1 A set of categories for aiding the identification of seismic suitability for well tie and interpretation

Acquisition	Processing	Visual	Analytical
Source and receivers	De-ghosting	Footprint penetration	Frequency range
Broadband	Q-compensation	Multiples	Signal-to-noise ratio
Maximum offset	Migration technique	Amplitude variations	Resolution
Fold map	Velocity model construction	Artefacts (linear)	Tuning
Aperture	FWI		
Water depth	Frequency balancing		
Line spacing			

incident angle incremental ranges for each of the stacks. This decision is only relevant where the geoscientist has access to the final post-migration gathers with any flattening and de-multiple processing applied (Chap. 4), with which to produce their own partial-stacks. It is common for the geoscientist to only have access to partial-stacks, however, access to the gathers from processing companies is usually possible. If gathers are available a few important considerations should go into deciding which increments and ranges to use. For these examples we will refer to angle stacks, but it is also relevant to offset stacks.

If the seismic interval of interest has an associated AVO effect, which is believed to be a possible indicator of the presence of hydrocarbon fluid content (DHI, Chap. 6), this is a consideration in the partial-stack angle range choice for well ties for both interpretation and seismic inversion. For instance, a Class 3 AVO response will exhibit an increase in seismic amplitude with increasing angle of incident across seismic gathers and partial stacks. So, even if the maximum amplitude response of the Class 3 AVO is at 45° we can probably see enough of an anomaly response at lower angles like 30°. At partial-stack angles approaching 45° data quality may not be optimum and show signs of residual moveout or have the dangers of critical angle associated with it. If, on the other hand, we are looking at a Class 2 AVO response with a particularly weak amplitude increase, due to geology or fluid properties, then it might be difficult to see a convincing anomalous amplitude change until we are out at +35°, in which case these larger angles need incorporating into our partial stacks.

The mute applied to the gather data can be quite subjective prior to stacking and is rarely performed by the interpreting geoscientist. The mute might be applied relatively aggressively to the gather data leaving no data past a certain angle at the depth of interest, thereby restricting the maximum angle range. Inversely, if the mute is very mild then the very far angle data might be good quality in some CDP locations but contaminated with poor data in other places. Including poor quality data into the partial stacks at the depth of the interval of interest will make achieving good well ties and wavelet estimation more difficult. The overlying of applied angle corridors on gathers at a number of key CDP locations can give the geoscientist an idea of the variability of maximum angle ranges at the target depth to help identify a sensible maximum partial angle stack range.

Near-angle traces (0°–3°) tend to be susceptible to noise artificially increasing true seismic amplitudes, making them 'hot traces'. If this is the case then including the first few angle degrees in a partial-stack can lead to the stack being of particularly poor quality, making well tying very challenging. Additionally, it should be noted that a near-angle stack with this issue is especially damaging for use in both a single-stack and multi-stack seismic inversion, as well as having a large negative effect on the P-impedance product which relies on near-angle stack reflectivity for its calculation. If noise attenuation methods are unable to suitably deal with the noise problem then a simple solution to noisy near traces is to simply drop them from the partial-stack range.

5.2 Wavelet Estimation and Well Tie

We know that performing the task of wavelet estimation varies in timescale and complexity depending on factors relating to data (wells and seismic) quality and availability along with the subsurface geology. Added to which all geoscience software packages have slightly different workflows with varying options. As such, this section is not meant as a step-by-step guide to performing wavelet estimation and well tie, but rather as guidance on the application of different wavelet methods and how they are used in the well-tying process. General good practice is to start with a theoretical wavelet for an initial well

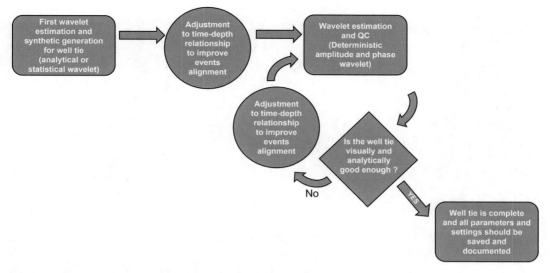

Fig. 5.6   A general workflow for wavelet estimation and well ties using an iterative workflow

tie and then to refine the well tie through an iterative process of adjustments. This workflow is shown in Fig. 5.6.

The purpose of the first wavelet is to allow the synthetic traces to be generated for a preliminary evaluation of how well the checkshot or VSP time-depth relationship ties the well and the seismic data. Well markers and pre-existing seismic interpretation maybe useful in this initial process. It is incorrect to assume that by using a checkshot or VSP the time-depth relationship will be correct with no uncertainty and subsequent interpretation will just be a formality. This first well tie also provides a chance to identify any missed data errors in the well-logs (e.g., gaps, spikes). The most common methods to generate this initial wavelet are statistical and analytical.

The geoscientist has a number of ways to estimate the wavelet for the well tie using modern geoscience software tools. Below are some standard methods for obtaining well-tie wavelets in order of least to most accurate:

(a) Select a standardized, analytical wavelet from a software library and define the frequency content (which makes the assumption that a known phase is constant);

(b) Estimate a wavelet from the seismic data statistically representing the amplitude and frequency (no phase estimation);

(c) Estimate the amplitude and phase of the wavelet from the seismic data.

5.2.1   Analytical Wavelets

The starting place for a wavelet used in a well tie is often analytical. These wavelets are universal in their application and require no seismic data to generate. Standardized mathematical functions control their characteristics although the user can input certain parameters such as frequency content, polarity and length. As these wavelets require no complex estimation workflow they are used ubiquitously for well ties. It is important to understand the variations of this wavelet class as some are more suited to well ties and forward modelling with modern broadband data than others. It should however be noted that these wavelets do not represent the character of the seismic data the well is being tied to and should therefore be used for the initial well tie with a view to estimating a more complex wavelet in

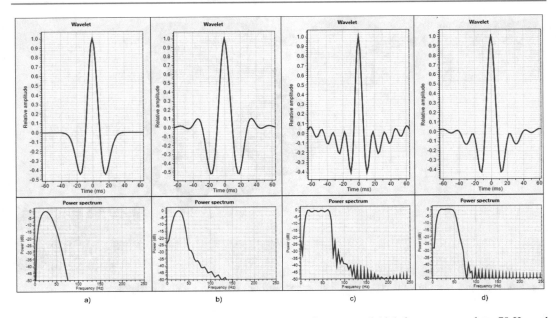

Fig. 5.7 Amplitude (top row) and frequency spectra (bottom row) for a: **a** Ricker wavelet with 2 ms sampling rate and central frequency 25 Hz; **b** Ormsby wavelet with 2 ms sampling rate and $f_1 = 10$ Hz, $f_2 = 20$ Hz, $f_3 = 35$ Hz and $f_4 = 40$ Hz; **c** Klauder wavelet with 2 ms sampling rate and high-frequency equal to 70 Hz and low-frequency equal to 10 Hz; **d** Butterworth wavelet with 2 ms sampling rate, low cut at 10 Hz with a low slope of 18 dB/Octave, a high cut of 50 Hz and high slope of 72 dB/Octave

further iterations, as shown in the well-tie workflow in (Fig. 5.6).

The most used analytical wavelet for seismic well ties is perhaps the Ricker wavelet, Fig. 5.7a. A zero-phase wavelet has two sidelobes is mathematically defined as the second derivative of the Gaussian function. The Ricker wavelet can be defined with a single frequency value making it very easy to test a range of frequencies for modelling purposes. This preferred wavelet has an asymmetrical amplitude spectrum and the peak frequency is defined as that with the most energy. The sidelobes are simple, smooth shapes making it particularly user friendly when analysis involving destructive or constructive interference occurs in modelling. The Ricker wavelet in the time domain is mathematically defined as

$$r(\tau) = \left(1 - \frac{1}{2}\omega_p^2\tau^2\right)exp\left(-\frac{1}{4}\omega_p^2\tau^2\right), \quad (5.1)$$

where τ is the time and ω_p is the most energetic frequency. Using a Fourier transform the Ricker

wavelet can be expressed for the frequency domain:

$$R(\omega) = \frac{2\omega^2}{\sqrt{\pi\omega_p^3}}exp\left(-\frac{\omega^2}{\omega_p^2}\right). \quad (5.2)$$

The Ricker wavelet (Fig. 5.7a) is smooth in appearance and the frequency spectrum is quite similar to that seen on older seismic data (before broadband seismic data) giving a hill-like profile with a focus on the mid-amplitude. This frequency spectrum is not representative of modern broadband seismic data sets which have been de-ghosted to allow a flatter, longer top to the frequency spectrum along with highlighting the frequency content in the lows and highs (Fig. 5.7a).

Like the Ricker wavelet, the Ormsby wavelet (Fig. 5.7b) is also a zero-phase wavelet, however the sidelobes are more complex. Also noticeable about the Ormsby is the frequency spectrum that exhibits a flatter frequency top as opposed to the hill seen on the Ricker. It is trapezoidal in shape and controlled by the four frequency points that are defined by the user (e.g., 5–10–40–50 Hz). The frequencies are arranged as a low cut (f_1), low pass (f_2), high pass (f_3) and high cut (f_4). Due to the shape of the spectrum it is often referred to as a band-pass filter, because the sharp, almost vertical, edges of the frequency spectrum have few frequencies with significant amplitude below end points defining the flat top. The Ormsby is more flexible to matching the frequency of the seismic data and is a good choice for use with modern broadband data sets. The Ormsby wavelet in time is represented by

$$A(t) = \frac{\pi f_1^2}{f_2 - f_1} \sin^2(\pi f_1 t) - \frac{\pi f_2^2}{f_2 - f_1} \sin^2(\pi f_2 t)$$
$$- \frac{\pi f_3^2}{f_4 - f_3} \sin^2(\pi f_3 t) - \frac{\pi f_4^2}{f_4 - f_3} \sin^2(\pi f_4 t).$$
$$(5.3)$$

The next wavelet to consider is the Klauder wavelet (Fig. 5.7c). When compared to the Ricker and Ormsby wavelets the Klauder is most like the Ormsby, in that is has more sidelobe reverberations present than the cleaner Ricker. Also, the frequency spectrum is very similar to a trapezoid shape but with a slightly more rugose top. The similarity of all three wavelet forms is that they are all zero-phase. The Klauder wavelet is associated with the onshore Vibroseis acquisition methods being an autocorrelation of a linear swept frequency-modulated sinusoidal signal. The Klauder is represented by the following:

$$A(t) = Re \left[\frac{\sin(\pi k t (T - t))}{\pi k t e^{2\pi i f_0 t}} \right]. \quad (5.4)$$

Also, commonly available to the geoscientist in most modern software is the Butterworth wavelet (Fig. 5.7d). The Butterworth wavelet is fairly unique compared to the other wavelets in that it is a minimum-phase wavelet. This makes its apparent appearance very different being asymmetrical with a peak amplitude followed by a single sidelobe reverberation. Like the Ormsby, the Butterworth is defined by using the two beginning values which define the upper and

lower frequencies and then the two values to define the lower and upper cut values.

5.2.2 Statistical Wavelet

The statistical wavelet estimation approach uses statistical methods of time-series analysis from the seismic data. Statistical wavelets give a simple but data-driven initial wavelet for the well tie and remove some of the complexities associated with defining a deterministic wavelet. For a geoscientist who is not familiar with well ties or, in particular, projects where estimating a deterministic wavelet is challenging (e.g., short well data length, poor well-log responses, noisy seismic data) it allows a well seismic synthetic to be generated and well tie achieved with relative ease.

The statistical wavelet estimates representative amplitude and frequency properties for the wavelet from the seismic data but makes no estimation of seismic phase. Knowledge of the seismic data should be used to make sure the polarity and phase of the wavelet are correct. Any degree of phase rotation is possible to apply in modern software, however, the best practice is to keep making a zero-phase assumption at this stage. There is justification for this assumption as most modern seismic data are expected to be processed as zero-phase. Another reason to make this wavelet zero-phase is because it removes any rotation in the generated well seismic synthetics, which may limit the usability of the first well tie.

Modern software should give the geoscientist a method for selecting the location of the traces that go into the estimation process. For the statistical wavelet, this often involves defining a window or probe and can be focused around the well. If the window is too big and picks up traces with poor (seismic and log) data, highly heterogeneous geology or significantly variable seismic character, the result may be a non-representative amplitude and frequency spectrum. In addition, strong reflectors such as unconformities and areas heavy in low-frequency can dominate the wavelet, so it is worth testing a few windows to understand these effects. The window should be focused to the well you are trying to tie. Most methods work around the principle of taking all the seismic traces within the input area and performing a type of autocorrelation which is parameterized based on the selected wavelet length or interval of interest input by the user. The frequency is derived as average approximation, the spectra are then stacked together and the Fourier transform is performed to generate the wavelet. Finally, a generic smoother is also often applied to reduce reverberation (i.e., sidelobe) energy.

Once an initial wavelet is created it can be used to generate the synthetic seismic traces from the well, which is then compared with the recorded seismic traces. As a simple rule the well should be tied from top to bottom. The objective of the well tie will dictate the total length along the well which is required to be tied. Sometimes a short well tie within a discrete reservoir interval is all that is required; more regional studies may require longer sections of the wells to be tied. An example from a standard software used in the energy industry to extract wavelets and perform seismic-to-well tie is shown in Fig. 5.8.

There are several basic checks to carry out.

Visual:

(a) Does an increase in acoustic impedance match the correct polarity in the synthetics (i.e., peak or trough)? If not, check the polarity of your wavelet.

(b) Displaying the reflection coefficients are very helpful to observe contrasts and understand synthetic events' strength and position.

(c) Do the synthetic trace characters look similar and line up to the corresponding seismic events?

(d) If you already have existing seismic interpretation, you may display these in the well tie window to see how closely they match the corresponding well markers;

(e) If the input logs are particularly spiky or the hole condition is poor is this effecting the reflection coefficient and tie in a negative way?

(f) Lithology logs (i.e., Vshale) can be useful to understand changes in log response.

Analytical:

(a) Is the cross-correlation value between the seismic and synthetics higher than 0.7 over the interval of interest? If so, this may be considered a good enough tie.

(b) Do the sidelobes of your wavelet appear to cause any constructive or destructive inference or tuning?

Once the initial analysis of the first well tie has been carried out it may be apparent that some or all synthetic and seismic events appear vertically misaligned. If this is the case a time-depth correction can be applied in the form of a simple bulk shift up or down of the well data. This is a simple static shift to the time-depth relationship applied to the whole well. Depending on the quality of the data used for the initial time-depth relationship the shift may be in the order of a few seismic samples (8–12 ms) or much larger (20–30 ms). This bulk shift should be focused at aligning the events of interest only, as trying to align all the events in the well is unlikely to be possible, or required. Should a very large bulk shift be required (e.g., more than 100 ms) then a review and possible revision of the initial time-depth relationship should be made.

The initial wavelet estimation and well tie performed can be replicated for other wells. Should a good enough synthetic-seismic tie be observed, it is reasonable to begin interpretation for a quick first pass data appraisal or mapping exercise. If the geology is well understood and key reflectors of interest are well imaged in the seismic data, then this level of well tie will support a fit-for-purpose interpretation. For interpretation requiring a higher level of technical product, this level of wavelet estimation and well tie is insufficient. The geoscientist can improve the quality and accuracy of the well tie by continuing to the next phase of wavelet estimation. The use of the deterministic method is suggested for two main reasons:

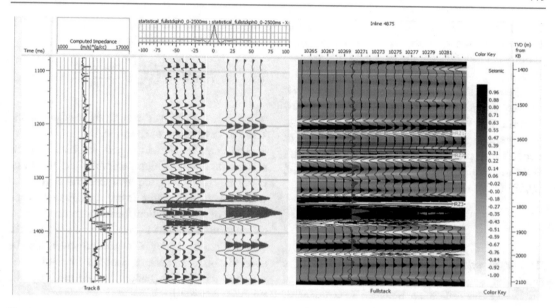

Fig. 5.8 Seismic-to-well tie from a real application example. The left plot shows a P-impedance log. The middle plot compares the observed (blue) and synthetic (red) seismic. The right plot shows the synthetic trace plotted in red overlying the observed seismic

(i) The uniform and symmetrical shape of the statistical wavelet, especially its sidelobes, will not reproduce the true character of the seismic wavelet in the well synthetics. Subtle seismic reflections and character variation will be overprinted, which could be critical to a difficult interpretation;

(ii) The phase of the wavelet is not truly captured with the statistical wavelet and this is an important factor to understand as it will ultimately underpin how and where geological boundaries are picked on the seismic data.

5.2.3 Deterministic Wavelet

The deterministic method is different to the statistical method in that it uses a direct measurement of both the seismic wavefield and measured well-log data to understand conditions at interfaces. When carrying out a deterministic wavelet estimation we are seeking a greater understanding of the embedded wavelet in the seismic data

than can be achieved with the statistical wavelet; an estimate of not only the amplitude and frequency content of the seismic data but also an estimate of the phase. The method commonly used in industry for deterministic wavelet estimation is that of White (1980) and White and Simm (2003). The well and seismic data types are correlated with coherency matching which uses a least-squares fitting technique to estimate the best wavelet from the seismic data. However, to be effective the well-logs must be of good quality and the initial time-depth relationship should give a reasonable well tie. This method is generally favoured above others because it provides a number of useful statistical measurements which help to define the quality/accuracy of the well tie, in what is termed 'goodness-of-fit'.

Goodness-of-fit is useful because relying on only the cross-correlation coefficient between the synthetic seismogram from the well and the observed seismic trace can be slightly deceptive. It relies on measuring the match of energy between seismic and synthetic peaks and troughs. Cross-correlation can be easily dominated by a single high-amplitude event with high correlation in a relative segment of data. There-fore, the window chosen for cross-correlation should be set to include a range of seismic events.

There is clear distinction to be made between the visual appearance of a good well tie and how accurate it is from an analytical peak and trough measurement perspective. The proportion of trace energy predicted by the matched synthetics is referred to as Percentage of Energy Predicted (*PEP*) (Simm and Bacon 2014)

$$PEP = 1 - \left(\frac{residual\ energy}{trace\ energy} \right). \quad (5.5)$$

The trace energy is the sum of squares of a particular segment of the trace. The residual energy is the difference between a segment of seismic and the corresponding synthetic seismo-gram from the well.

The *PEP* is usually presented in a map format which gives the geoscientist a spatial under-standing of the best match location of the syn-thetic to seismic tie in a user-defined search area around the well. The size of the window around the well does influence the estimated wavelet due to the inclusion of more trace for the *PEP* cal-culation. There is usually a software option to compare the estimated wavelet at the well loca-tion and the wavelet from the optimum *PEP* location within the pre-defined area. Interest-ingly, this is a great strength of the method because it acknowledges the uncertainty of the positioning of the seismic traces due to migration and the averaging caused by processes such as binning. It also acknowledges that the well position may not be accurate and the possibility that the well may have been drilled on or near a fault or geological anomaly. If this is the case trying to estimate a wavelet at the well will never be optimum and looking at seismic traces ~25 m away from the well position may give a more representative wavelet estimation.

The *PEP* is not the only way to measure the accuracy of the well tie. The normalized mean square error (*NMSE*) is a measure of the differ-ence between the synthetic energy errors when matching to the seismic and total energy of the

synthetic trace. The calculation of the *NMSE* is defined by the equation below:

$$NMSE \approx \left(\frac{1}{bT} \right) \left[\frac{1 - PEP}{PEP} \right], \quad (5.6)$$

where b is defined as the analysis bandwidth and T is the data segment duration.

The *NMSE* calculation will be biased if the seismic data have a high percentage of noise that is not represented in the synthetic trace from the well. It can also suffer from inaccuracies when the size of the analytical bandwidth used in the calculation is either too narrow or broad for the wavelet length. The *NMSE* can, however, also be used to obtain an idea of phase error seen in the wavelet within the bandwidth defined by the seismic data used:

$$E\{phase\} \approx \sqrt{\left(\frac{NMSE}{2} \right)} (in\ radians). \quad (5.7)$$

Looking at both the *PEP* and *NMSE* will give a quantitative guide to the quality of the wavelet estimation. Guidance on the numerical threshold of these measures to be reached in order to identify a "good" well tie is given in some publications as *PEP* > 0.7 and *NMSE* < 0.1 (Simm and Bacon 2014). However, when a well tie fails to meet or exceed these defined thresh-olds it does not mean the well should be dis-carded. Ultimately it is for the geoscientist to decide how best to manage this uncertainty and derive any value from a sub-optimal well tie.

5.2.4 Wavelet Length

The length of the wavelet used for the well-tie process varies based on the discretion of the interpreter, though there are a few principles that are worth considering. The wavelet length needs to be long enough to capture the discrete wave-form you are trying to estimate. Most wavelet estimation tools give the option to apply some sort of taper (e.g., cosine or Papoulis) to the wavelets to dampen the reverberations associated with the sidelobe energy. The wavelet must, therefore, be long enough for the tapers to not affect the main peak or trough. The wavelet length should be driven by the tie window and tie itself. Statistics such as goodness-of-fit and matching can be used to identify and optimize length. Phase and appearance of the wavelet also changes with wavelet length and helps in deciding the optimum length. The length of well data that is available for the wavelet estimation should also be considered. The wavelet estimate window should be approximately three to five times the length of the wavelet for a 4 ms sampling rate. So, a 120 ms wavelet requires an estimation window of 360–600 ms and for a 200 ms wavelet it would be 600–1000 ms. Larger time windows (1000 ms) can be prone to exhibit variability in amplitude, frequency and phase as a function of attenuation and the filtering effects of the earth. Strong variations in geology within the well tie window can also cause these effects. It is unrealistic to expect to tie every event in a 1000 ms TWT window and so the focus should be on the main events of interest.

When considering estimating wavelets for the purpose of seismic inversion we want to capture any significant variability of the wavelet properties throughout the interval of interest. If the data are variable in amplitude, frequency and phase, vertically/time varying wavelets can be used. It is also possible to create horizontally varying wavelets to cope with large lateral changes in either geology or seismic quality. The purpose of these wavelets is to capture the variation in the seismic data so that the deconvolution, which is part of the inversion process, is performed using a relevant wavelet. These wavelets can be very sophisticated even incorporating estimations of attenuation to properly understand variations in amplitudes.

5.2.5 Stretch and Squeeze

The time-depth relationship may require changing by bulk shifting the well in order to improve the alignment of peak and trough events between the seismic and the synthetics. While this level of adjustment is generally all that is required for a well tie to be suitable for interpretation in a seismic inversion study, an additional step can be applied which improves the tie even further but it requires careful consideration.

The data used for the initial time-depth relationship may contain localized inaccuracies in velocity over some intervals or boundaries which do not correctly capture the true velocity of the geology (e.g., when using a time-depth relationship for a well with significantly varying geology and velocity). These inaccuracies lead localized event alignment mismatches which cannot be fixed by moving the whole well with a bulk shift, but can be corrected by selecting the zone and changing the velocity to be either slower or faster, to align the reflectors/events. This method is known as stretch and squeeze.

The quality of the checkshot or VSP at the start of the process will have been reviewed. It is common to remove knee points from a checkshot to give a smooth velocity profile. This stage in the process may need to be revisited if the stretch and squeeze is required in multiple locations down the well. If no checkshot or VSP was available and a sonic log or extracted log from a velocity model was used for the initial time-depth model then the application for stretch and squeeze may be more valid.

The workflow targets a particular part of the time-depth relationship which must be either slowed down (i.e., stretched) or sped up (i.e., squeezed). The smaller the interval selected the more extreme the effect will be, so it is advisable to pick a larger interval in order that the adjustment is not so severe. The most important measurement in this workflow is the drift curve which shows and applies variations to the time-depth curve away from zero as a percentage of velocity. Should the applied correction alter the velocity of the by more than 10–15% this is considered excessive and should be reduced to make the best match possible. It is also important to think about velocity alterations in terms of geology and decide if values seem plausible for the rock type and particular depth. If the well is to be used as input data into the generation of a starting model or low-frequency model for an inversion then going beyond the 10–15% velocity change may be acceptable. However, the impact of the miss-tie should be reviewed especially if it occurs at an event matching a key framework horizon event.

5.3 Practical Guide to Well Ties and Wavelet Estimation

When performing seismic-to-well tie and wavelet estimation the available, or selected, well for ties must fall within the spatial coverage of the available seismic data. This might look obvious but in difficult settings with large amounts of data acquired in different periods might not be straightforward.

After the well is selected, we need to look at the available well-log suite as well-logs are a key input data for well ties. A full petrophysical review of well-logs ensures the best possible quality of well ties is achieved. Finally, from the existing data set the presence of well markers is critical and should be consistent across wells and be suitable for the interpretation goals. After reviewing the available data in terms of numbers of well, well tops and log suites, the initial time-depth relationships should be robust so that the first well tie is of decent quality.

As a final message and as best practice, the time and complexity spent on the wavelet estimation should be proportional to the technical level and importance of the following step of the geo-modelling workflow. These steps comprise conventional seismic interpretation and advanced modelling methods such as seismic inversion.

References

Feng F, Grana D, Balling N (2021) Imputation of missing well log data by random forest and its uncertainty analysis. Comput Geosci 152:104763

Herrera RH, Fomel S, van der Baan M (2014) Automatic approaches for seismic to well tying. Interpretation 2: SD9–SD17

Jilinski P, (2015) Borehole log guided seismic spectral enhancement applied for well to seismic tie workflow. SEG Global Meeting Abstracts: 1117–1120.

Lindsay R, Koughnet RV (2001) Sequential Backus averaging: upscaling well logs to seismic wavelengths. Lead Edge 20(2):188–191. https://doi.org/10.1190/1.1438908

Nivlet P, Smith R, AlBinHassan N (2020) Automated well-to-seismic tie using deep neural networks. SEG Technical Program Expanded Abstracts: 2156–2160.

Simm R, Bacon M (2014) Seismic amplitude: an interpreter's handbook. Cambridge University Press

White RE (1980) Partial coherence matching of synthetic seismograms with seismic traces. Geophys Prospect 28:333–358

White RE, Simm RW (2003) Tutorial—good practice in well ties. First Break 21:75–83

Interpreting Seismic Amplitudes

This chapter focuses on defining AVO and exploring the subsurface conditions and data required to observe the phenomenon. We look at how AVO can be used to identify variations in rock properties along with the presence of different pore-fluids in the subsurface. Throughout the chapter the many assumptions and pitfalls associated with AVO are discussed from a practical perspective. We will end with a summarized guide for the geoscientist to use when embarking on AVO analysis.

It is important to note that the terms AVO and AVA refer to amplitude anomalies in either offset (AVO) or angle data (AVA). However, practitioners in the energy industry rarely adhere to this and instead use the term AVO to cover both data domains. In keeping with industry practices this book simply refers to AVO to cover both data domains.

6.1 AVO Anomalies

6.1.1 Seismic Interpretation and Amplitude Maps

Seismic interpretation within the energy industry is mainly concerned with mapping reflectivity events that represent either a soft (i.e., I_P decreases) or hard (i.e., I_P increases) boundary or interface between two geological units. The interfaces are represented in the seismic reflection data as either peaks or troughs depending on the polarity of the data. The amplitude of the peaks and trough is defined by the reflection coefficient which quantifies the proportion of seismic wave energy reflected from an interface as a wave is incident on it. The reflectivity peaks and troughs are tied to well-logs using the common workflow of seismic well tying (Chap. 5), which enables the seismic events to be correlated to known geological units and boundaries. What follows for the interpreter is to carry out mapping of key events away from the well control by tracing or following a particular peak, trough or even zero crossing throughout the seismic volume. The result will be a three-dimensional surface such as the top or base of a geological unit. This activity allows discrete stratigraphic units to be mapped over small, focused areas such as a prospect or license block, as well as larger regional areas for the purpose of activities such as gross depositional environment (GDE) mapping. Seismic interpretation is usually carried out on a full-stack seismic data set because the characteristics of good signal-to-noise properties and broad frequency content are present. Figure 6.1 shows an example of a seismic section from a field in the Norwegian North Sea with a horizon interpretation on the left and a final interpreted three-dimensional surface on the right.

Most interpreters favour seismic data displayed in density colour rather than wiggles. With this display setting the stronger the colour of the event the stronger the amplitude, which is

T. Tylor-Jones and L. Azevedo, *A Practical Guide to Seismic Reservoir Characterization*,
Advances in Oil and Gas Exploration & Production, https://doi.org/10.1007/978-3-030-99854-7_6

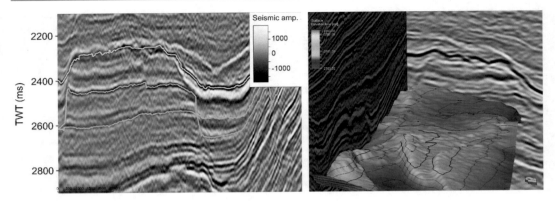

Fig. 6.1 Vertical seismic section with multiple horizons picked on a seismic volume (left) and the resulting three-dimensional horizon for one of the surfaces (right)

indicative of a larger contrast at the interface and a bigger reflection coefficient—be it positive or negative. Mapping a seismic event with a strong and consistent amplitude response is far easier than a weak or variable response, yet these amplitude variations can communicate a lot more about the rock properties beyond whether the lower layer is acoustically harder or softer.

Though detailed interpretation can be time consuming great value can be added by providing the interpreter with a unique knowledge of the amplitude variability and strength in the data. Variability of seismic amplitude strength for a specific event is often observed by an experienced interpreter while mapping a prospect area or during larger regional mapping exercises. Increased exposure to a diversity in the quality of seismic data as well as different basins and geological settings allows the interpreter to become more aware of subtle features such as amplitude behaviour. A large part of interpretation experience comes from pattern recognition. Successfully developed (economic hydrocarbon production) seismic features identified in past mapping can be searched for and those that led to failure can be avoided (Chopra and Marfurt 2005).

Seismic amplitudes provide the interpreter with some qualitative information about the rock properties. However, a sense-check filter should always be applied to the seismic quality and suitability during any amplitude analysis. As discussed in Chap. 2, the seismic experiment has

many uncertainties that we try to understand in the processing workflow to produce the best image of the subsurface. Below are some key points to keep in mind:

(a) Acquisition type and parameters ultimately define the limits of what is possible to image, even with the best processing tools;

(b) Processing workflows can have a dramatic effect on the true amplitude of events;

(c) Migration is an approximate solution for positioning reflectors (especially complex dipping geology) and is not the absolute answer.

The preservation of true seismic amplitudes within seismic data is often brought up when amplitude variation is identified during interpretation. In simple terms, true amplitude in the seismic data reflects the true contrast behaviour and magnitude of the rocks being imaged. All processing projects should safeguard true amplitude preservation, which lies with the processor(s) being mindful of how each seismic processing step applied to the data affects the strength and spatial variability of amplitudes. One commonly implemented QC is to carry out periodic migrations throughout the processing workflow to check the effects of pre-migration parametrization on the stacked image, since interpretation is mainly done on stacked data. During this QC integration, geological understanding is very important in helping to guide decision-making.

6.1.2 Why is Amplitude Variation Useful?

If observations are made during the seismic interpretation process that show evidence of areas of amplitude variations, then further investigation is often warranted. The interpretation surface picked on a peak or trough event (Fig. 6.1) allows the geoscientist to make an amplitude extraction map showing the variations of the amplitudes spatially (Fig. 6.2). This is one of the biggest values of having three-dimensional seismic data over two-dimensional seismic data, especially for exploration. The type of geology the horizon is used to map will determine what insights the amplitude map may give on the area of interest. If the horizon is used to map a reservoir unit, then discrete sedimentary units and edges might be revealed; however, if the horizon is used to map a top-seal formation then low variability on the amplitude map would suggest a uniform-seal unit. This is also where the accuracy of the interpretation comes into play, although most amplitude maps are extracted from a window around the horizon.

The purpose of mapping amplitude variations in this way is to provide geophysical evidence to support the geological, sedimentary and geomorphological models. The type of project setting (e.g., exploration, appraisal or development) will give additional focus to the key objectives for which an amplitude map may be used. Irrespective of the type of project three broad categories of information are always of interest to the geoscientist:

(a) Depositional processes/environments—aeolian, fluvial, lacustrine, marine;
(b) Rock properties—reservoir presence, reservoir quality;
(c) Fluid properties—water, hydrocarbon fluids.

Figure 6.2 illustrates three different amplitude maps showing unique depositional features that are defined by higher seismic amplitudes compared to the general background trend. Analysis of maps like these would allow an interpreter to draw some conclusions, for example about the depositional environment, distribution of lithologies, edges of individual systems.

Reservoir presence and rock properties can be inferred directly from amplitude maps such as those shown in Fig. 6.2. Visual observations such as the spatial distribution of the strongest and weakest amplitudes within a feature are often very insightful and relatable to the spatial distribution of certain geological formations. However, further analysis and integration with geological models, local analogs and known well penetrations of similar features will be required before drawing further quantitative conclusions. Although the amplitudes in these features appear convincing, cases of over corrected/processed amplitudes within the seismic data have been known to make these features change dramatically.

Fluid properties are the most difficult to infer from a simple amplitude map. As such, the map should be viewed as a signpost to suggest the amplitude variability in the seismic data is responding in a way consistent with the recognizable presence of hydrocarbons. To get a better understanding of the response for fluid properties the interpreter needs to consider amplitude anomalies and more specifically AVO.

6.1.3 What is a Seismic Amplitude Anomaly?

An amplitude anomaly is a seismic amplitude that deviates from what is normal or expected. Amplitudes are expected to show minor variations in strength along a particular reflector without changing geology, in the order of $\sim 5\%$–10%, however an amplitude anomaly stands out in the range of 20%–50% (or greater) difference in strength. An anomalously high-amplitude response suspected of indicating the presence of hydrocarbons is commonly known as a bright spot. The assessment of seismic quality and acquisition suitability, mentioned earlier in this chapter, should play a part in helping the interpreter calibrate the characteristics of seismic amplitude variability in a particular seismic data set. When an amplitude anomaly is identified, the interpreter should try to remain unbiased as to its

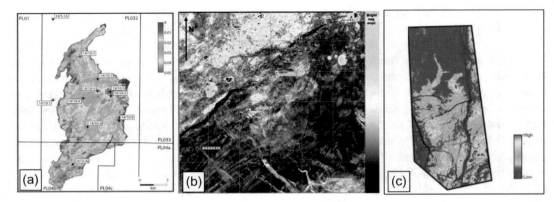

Fig. 6.2 Three amplitude maps display how seismic amplitudes show depositional features. **a** Fan reservoir system in the Sea Lion Field, North Falkland basin (Bunt 2015); **b** Bright amplitudes show sand channels and sand ridges on the fluvial-deltaic plain background offshore Brazil (Klarner et al. 2008); **c** Channel systems with tributaries and fan region Farsund Basin offshore Norway (Phillips et al. 2019)

meaning, which at the time of identification is likely unknown and speculative at best. Even new amplitude anomalies identified in mature areas where active developments produce from "similar looking" amplitude(s) anomalies should be treated in this way.

To understand amplitude anomalies in seismic reflectivity data the geoscientist extracts a number of derivatives to provide information about the amplitude properties. This step commonly includes looking at amplitude variations beyond the simple full-stack input data. AVO attributes are specifically concerned with understanding the seismic amplitude in the context of its ability to indicate the presence of hydrocarbon fill and to distinguish against non-hydrocarbons. It is worth noting that a distinction should be made between focused seismic AVO attributes and the generic term seismic attributes, which is a broader term that encompasses a huge range of derivatives from seismic data (i.e., calculations performed over the amplitude, frequency and phase content of the seismic signal). Although a large majority of the seismic attributes measure and use amplitudes strength and continuity as input for algorithms, they are also used to help the interpreter in fault detection, structural features identification and thickness calculations.

There are different definitions for seismic amplitudes that indicate the presence of fluids in the subsurface. Below are some common examples with basic definitions:

- Bright spot—seismic amplitudes that are much stronger than those surrounding them. On a density colour or wiggle trace display of seismic data, they will appear brighter in colour than those around them or of the same geological unit. They will have a corresponding high-amplitude value.
- Dim spot—seismic amplitudes that are much weaker than those surrounding them. On a density colour display or wiggle trace display of seismic data will appear dimmer in colour than those around them or of the same geological unit. They will have a corresponding low-amplitude value.
- Flat spot—if a seismic reflector exhibits a ruler flat horizontal line over a number of traces and is below the top of the reservoir reflector this is called a flat spot and may indicate a fluid contact (i.e., oil above water or gas above oil).
- Direct Hydrocarbon Indicators (DHIs)—an amplitude (or attribute) response caused by the presence of hydrocarbons and shows their spatial distribution in the subsurface. This term is used in the industry to describe general amplitude anomalies often without the rigorous work done to back it up. It should be used

in the strictest terms where enough investigation has been carried out with available data to rule out other potential causes.

- Direct Detection Indicators (DDIs)—same as DHI, meaning the seismic signature is directly coming from the presence of hydrocarbons.
- Hydrocarbon Indicators (HCI)—same a DHI and DDI.
- AVO—Amplitude Versus Offset or sometimes referred to as Amplitude Variation with Offset is a general term for describing how a seismic amplitude changes its characteristics such as strength, shape, phase and polarity with changes in distance between the source and receiver (offset) for a single seismic peak or trough.
- AVA—Amplitude Versus Angle or sometimes referred to as Amplitude Variation with Angle is a general term similar to AVO except rather than using offset distance from the source and receiver, we refer to an angle corridor derived from offset domain using a seismic velocity field.

6.1.4   Brief History of Bright Spot and Early Origins of AVO

The use of bright spot interpretation comes from a time when most exploration focused on mapping and drilling structural features (e.g., structural highs) seen on seismic data. The resolution of early seismic data was often very poor and the ability to correctly image dipping reflectors was basic. However large anti-cline features could be identified and often proved to be excellent traps for large accumulations of hydrocarbons. It is worth remembering that most pre-1970s seismic data was two-dimensional and viewed in an unmigrated (or poorly migrated) state like the example, taken from the North Sea, shown in Fig. 6.3.

Before the advent of digital data, seismic data was only available in analog which limited the accuracy of recovering true amplitudes, thereby restricting any useful application of seismic amplitude variation. Evidence shows that during the 1950s and 1960s seismic attributes (e.g., structural elevation, dip, thicknesses and fault discontinuities) were routinely used without digital seismic data or modern computers (Chopra and Marfurt 2005). There is also some evidence that in Russia the use of amplitudes was being applied to help select structures and identify hydrocarbon contacts (Churlin and Sergeyev 1963; Hilterman 2001).

By the late 1960s and early 1970s publications show an increasing acceptance of the use of bright spots for hydrocarbon identification. In 1973 the Geophysical Society of Houston sponsored a symposium on bright spots and other seismic attributes linked to the indication of potential hydrocarbon reserve detection (Hilterman 1975). There is also a mention in a 1976 publication by Levin et al. (1976), which reviewed developments in geophysics from 1969 to 1974, that the DHI was becoming more common in geophysical vocabularies.

Bright spot analysis was a step forward, but it was focused on interpretating post-stack seismic data and although it proved tremendously useful to support the exploration and drilling of many prospects it was by no means fool proof. Many dry holes were still drilled on targets supported by bright spots. It is, however, worth noting that even today some exploration companies use bright spots on post-stack data as DHIs.

It is important to make a clear distinction between the use of bright spot amplitude analysis in post-stack data and the move into pre-stack data analysis, which is the domain of AVO. It is widely recognized that the value of pre-stack amplitude variation analysis was brought to the attention of the industry by Ostrander (1982), who recognized the usefulness of AVO in increasing success of hydrocarbon discoveries but also noted that "many seismic amplitude anomalies are not caused by gas accumulation". Ostrander proposed looking at Poisson's ratio to understand how the reflection coefficient of materials with different Poisson's ratios varied with the angle of incident of seismic waves (Fig. 6.4).

Fig. 6.3 A vintage (1968), unmigrated seismic image from the North Sea with some structural features visible (Paulson and Merdler 1968)

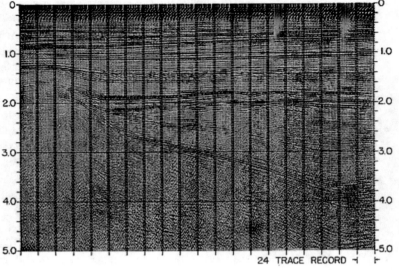

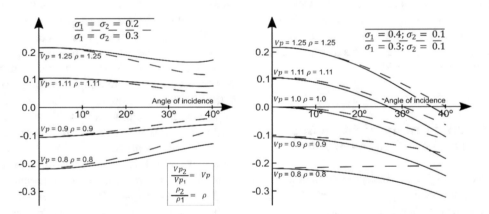

Fig. 6.4 Plot of P-wave reflection coefficient versus angle of incidence for constant Poisson's ratio of 0.2 and 0.3 (left) and reduction in Poisson's ratio (right) (Ostrander 1984)

Ostrander's work (1984) brought the concept and application of AVO to the broader industry that had, until then, relied on bright spots. Yet in the 1950s and 1960s, numerous publications recognized the importance of elastic rock properties and the effects of wave splitting and conversion of P-waves to S-waves due to varying incident angle. Most of the published works focused on practical laboratory experiments and results to prove these concepts; a clear indication that it was before the time of easy access to computer modelling resources (Mann and Fatt 1960; Clay and McNeil 1955; Wyllie et al. 1956). By the 1980s and early 1990s the increasing sophistication of computers and the advances and commonality of three-dimensional seismic surveys led to further uptake in amplitude analysis. As more successful wells were

Table 6.1 Table showing the major milestones in the evolution of AVO understanding (Hilterman 2001)

Year	Author	Method
1899	Knott	Amplitude versus Incident Angle
1919	Zoeppritz	Amplitude versus Incident Angle
1951	Gassmann	Petrophysical link to seismic
1955	Koefoed	Poisson's ration from RC(θ)
1961	Borfeld	Linear approximation
1976	Rosa	RC(θ) elastic inversion
1984	Ostrander	AVO integration
1985	Shuey	Rock properties at variable incident angles

drilled following AVO interpretation and large discoveries were made in high-profile global basins, the uptake of the advanced techniques increased. Going beyond the early 1990s an almost exponential increase in the application of AVO can be seen, as well as several new techniques and workflows that remain current today. Table 6.1 shows the early historic progression of the understanding of AVO.

6.1.5 Why Do We Get AVO?

Analysis of AVO looks at seismic data in the offset domain, which has been through a pre-stack common-offset migration (AVO is used for dealing with data in the angle domain, where each trace on a seismic gather refers to a different incidence angle). Chapter 2.2.1 describes how wavefronts propagate out from a single source origin and are incident on reflection boundaries caused by changes in geology. It was also discussed that the down going wave cannot be assumed to be perfectly vertical (normal incidence angle) when hitting the interface event. Snell's law (Eq. 6.1) shows that a relationship exists between the incident angle of P-waves and the angle of reflected and refracted P-waves. Snell's law also acknowledges that S-waves are generated by splitting P-waves:

$$p = \frac{\sin \theta_1}{V_{P1}} = \frac{\sin \theta_1}{V_{P1}} = \frac{\sin \phi_1}{V_{S1}} = \frac{\sin \phi_1}{V_{S1}}. \quad (6.1)$$

As the P-wave hits a boundary between two elastic media layers, reflected and transmitted P-waves are produced along with a phenomenon known as wave splitting or mode conversion, which leads to the generation of reflected and transmitted S-waves. The amplitudes of the reflected and refracted P-waves (R_{PP}) and S-waves (R_{PS}) are observed to vary in relation to the angle of incidence of the down-going P-wave when it hits an interface event. This variation in amplitude with incident angle is the basis for AVO analysis. Figure 6.5 shows how the relative energy of different wave components, described by Snell's diagram, vary as a function of angle of incidence.

The Zoeppritz equations[1](1919) describe the partitioning on seismic wave energy for reflected and transmitted P-waves and S-waves caused by the wave being incident on a plane interface between two layers of elastic media. This is similar to the diagram in Fig. 6.5 which shows the partitioning of wave energy at the interface. The method considers the wave velocity along with the dilation effects of P-waves and the dilation-free effect of S-waves on an isotropic media with relation to Lamé's constant and Poisson's ratio (Zoeppritz 1919). The equations use three properties of the media above and below the interface: density, P- and S-wave velocities. The equations consists of four unknowns in the form of reflected P-, reflected S-, transmitted P- and transmitted S-wave amplitude coefficients. The matrix-based equation

---

[1] named after the German geophysicist Karl Bernhard Zoepptriz (1881–1908).

Fig. 6.5 Energy distribution for reflected and refracted P- and S-waves with varying source receiver angles (Dobrin 1976)

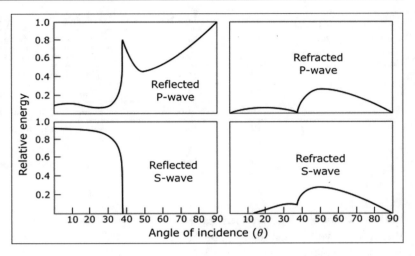

gives coefficients as a function of angle which can then give amplitude magnitude. The equation assumes the incident angle at the boundary is below critical angle. Critical angle is defined as the angle at which the refracted wave would propagate along the boundary of the interface and not give a true or expected ray path through the subsurface. For the analysis of AVO the most useful reflection coefficients are likely to be the R_{PP} (i.e., P-wave incident and P-wave reflected) and R_{PS} (i.e., P-wave incident wave and P-wave and split shear-wave reflected):

$$
\begin{bmatrix} R_P \\ R_S \\ T_P \\ T_S \end{bmatrix} = \begin{bmatrix} -\sin\theta_1 & -\cos\phi_1 & \sin\theta_2 & \cos 2\phi_2 \\ \cos\theta_1 & -\sin\phi_1 & \cos\theta_2 & -\sin\phi_2 \\ \sin 2\theta_1 & \frac{V_{P_1}}{V_{S_1}}\cos 2\phi_1 & \frac{\rho_2 V_{S_2}^2 V_{P_1}}{\rho_1 V_{S_1}^2 V_{P_2}}\cos 2\phi_1 & \frac{\rho_2 V_{S_2} V_{P_1}}{\rho_1 V_{S_1}^2}\cos 2\phi_1 \\ -\cos 2\phi_1 & \frac{V_{S_1}}{V_{P_1}}\sin 2\phi_1 & \frac{\rho_2 V_{P_2}}{\rho_1 V_{P_1}}\cos 2\phi_1 & \frac{\rho_2 V_{S_2} V_{P_1}}{\rho_1 V_{P_1}}\sin 2\phi_1 \end{bmatrix} \begin{bmatrix} \sin\theta_1 \\ \cos\theta_1 \\ \sin 2\theta_1 \\ \cos 2\phi_1 \end{bmatrix}, \qquad (6.2)
$$

where R_P is the reflected P-wave, R_S the reflected S-wave, T_P is the transmitted P-wave, T_S is the transmitted S-wave.

The Zoeppritz equations allow a calculation of reflection amplitudes as a function of incident angle and is considered the most complete and thorough way to make this calculation when carrying out AVO analysis. However, the Zoeppritz equation can be overly complex and difficult to work with when trying to quantify the effect of small changes in rock properties on a reflection coefficient. Although the equations solve for the four unknowns (R_P, R_S, T_P and T_S), to understand the wave behaviour at the interface the results are difficult to relate directly to rock properties—unlike Poisson's ratio. Finally, producing curves from a simple one-dimensional model can easily be achieved using Zoeppritz, but applying the method to large three-dimensional seismic data sets for AVO attributes or seismic inversion calculations can be computationally intensive. Over the years, numerous simplifications (linearized approximation) for P-S reflection amplitudes have been developed (Table 6.2). Most mainstream geophysical software has the option of running both well-based AVO models and partial-stack seismic inversion with either the full Zoeppritz or a more modern (linearized) approximation.

Table 6.2 Summary of the commonly used approximations of the Zoeppritz equations.

Autor	Equation	Comments
Bortfeld (1961)	$$R(\theta_1) = \frac{1}{2}\ln\left(\frac{V_{p2}\rho_1 \cos\theta_1}{V_{p1}\rho_1 \cos\theta_2}\right) + \frac{\sin^2\theta_2}{V_{p1}}$$ $$\left(V_{s1}^2 - V_{s2}^2\right)\left[2 + \frac{\ln\left(\frac{\rho_2}{\rho_1}\right)}{\ln\left(\frac{V_{p2}}{V_{p1}}\right) - \ln\left(\frac{V_{p2}V_{s1}}{V_{p1}V_{s2}}\right)}\right]$$	Bortfeld's is the first linearized approximation for the Zoeppritz equation and predicts the reflection amplitude for different incident angles and offsets. Not widely used in AVO analysis due to not explicitly indicating the relationship between reflection amplitude and angle/offset dependency (Yilmaz 1987)
Aki-Richards (1980)	$$R(\theta) = a\frac{\Delta V_p}{V_p} + b\frac{\Delta\rho}{\rho} + c\frac{\Delta V_s}{V_s}$$ $$a = \frac{1}{2\cos^2\theta}, \; b = 0.5 - \left[2\left(\frac{V_s}{V_p}\right)^2 \sin^2\theta\right],$$ $$c = -4\left(\frac{V_s}{V_p}\right)^2 \sin^2\theta$$	Approximation of Bortfeld to give a much simpler three-term equation using P-velocity, density and S-velocity
Hilterman (1983)	$$R(\theta_1) = \frac{V_{p2}\rho_2 \cos\theta_1 - V_{p1}\rho_1 \cos\theta_2}{V_{p2}\rho_2 \cos\theta_1 + V_{p1}\rho_1 \cos\theta_2} + \left(\frac{\sin\theta_1}{V_{p1}}\right)$$ $$\left(V_{s1} + V_s2\right)\left[3(V_{s1} - V_s2) + \frac{2(V_{s2}\rho_1 - V_{s1}\rho_1)}{\rho_2 + \rho_1}\right]$$	A simplification of Bortfeld's equation that separates the reflection coefficient into the form of acoustic and elastic properties
Shuey (1985)	$$R(\theta_1) = R_p + \left[R_p A_o + \frac{\Delta\sigma}{1-\sigma^2}\right]\sin^2\theta$$ $$+ \frac{\Delta a}{2a}\left(\tan^2\theta - \sin^2\theta\right)$$ $$\sigma = \frac{(\sigma_1 + \sigma_2)}{2}, \Delta\sigma = (\sigma_2 - \sigma_2)$$ $$A_o = B - 2(1+B)\frac{1-2\sigma}{1-\sigma}$$ $$B = \frac{\frac{\Delta a}{a}}{\frac{\Delta a}{a} + \frac{\Delta\rho}{\rho}}$$	A modification of the Aki-Richards approximation. This introduces the variable of Poisson's ratio for better understanding of rock properties. It is common to see this with either two or three terms. However, with two-terms its ability to predict AVO behaviour accurately breaks down beyond an offset angle of >30 degrees
Fatti et al. (1994)	$$R_{pp}(\theta_i) = \frac{1}{2}\left(1 + \tan^2\theta\right)\frac{\Delta I_p}{I_p} - 4\left(\frac{V_{s0}}{V_{p0}}\right)^2 \sin^2\theta\frac{\Delta I_s}{I_s}$$ $$- \frac{1}{2}\left(\tan^2\theta - 4\left(\frac{V_{s0}}{V_{p0}}\right)^2 \sin^2\theta\right)\frac{\Delta\rho}{\rho}$$	The Fatti method contains a specific density term and is valid up to critical angle
Smith and Gidlow (2003)	$$R_{pp}(\theta_i) = \frac{1}{2}\left(\frac{\Delta V_p}{V_p} + \frac{\Delta\rho}{\rho}\right) - 2\left(\frac{V_{s0}}{V_{p0}}\right)^2$$ $$\sin^2\left(2\frac{\Delta V_s}{V_s} + \frac{\Delta\rho}{\rho}\right) + \frac{1}{2}\tan^2\theta\frac{\Delta V_p}{V_p}$$ $$\rho = aV_P^b = a\left(V_p\right)^{1/4}$$ $$\frac{\Delta\rho}{\rho} = b\frac{\Delta V_p}{V_p} = \frac{1}{4}\frac{\Delta V_p}{V_p}$$	An approximation based on the Aki and Richards' equation. This equation makes use of the relationship between density and P-wave velocity removing dependency on a discrete density term

6.2 AVO Classifications

The standard used to describe and classify amplitude behaviour in the energy industry comes from the seminal work of Rutherford and Williams (1989), which introduced three classes of AVO characteristics that can be exhibited by gas sand reflectors. Their work focuses on the importance of understanding the normal incident angle (R0 or intercept) reflection coefficient represented by reflector amplitude strength, and how the reflection coefficient and corresponding amplitude varies as the incident angle is increased (G or gradient) with additional AVO classes, as well as the introduction of projecting AVO reflections into AVO intercept and gradient space (Fig. 6.6). Although their paper focuses on gas responses it is worth noting that this AVO classification can also be applied to oil accumulations.

One common misunderstanding when applying the AVO classification shown above is the polarity of the user's data, which may appear to be trivial but is an important matter. Figure 6.6 uses SEG normal polarity, which is the convention created by the Society of Exploration Geophysicists (SEG) for the display of zero-phase seismic data. In this case, an increase in P-impedance is a positive value and a decrease is a negative value. For data with the reverse polarity a Class 1 response shows a negative event getting less negative.

6.2.1 Class 1

Class 1 AVO behaviour is characterized by a strong amplitude response for an increase in impedance across the interface seen at normal incident angle; with progressive increase of incident angle the amplitude decreases (i.e., dims) and becomes weaker. The term "dim spot" or "dimming" is used to describe this behaviour in a full-stack seismic volume and indicates that the reservoir with hydrocarbons present is harder

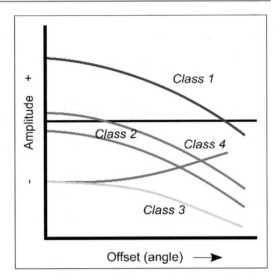

Fig. 6.6 The five modern (class 2 includes 2P) AVO classes used to describe how reflectivity amplitudes vary as a function offset or angle. It should be noted that the polarity in this image is SEG normal with a negative amplitude representing a decrease in impedance

than the overlying shale, but softer than the reservoir with brine. This is a difficult class to identify as anomalous from other amplitudes because, generally, amplitude behaviour can exhibit a Class 1 response as a function of seismic amplitude attenuation from the longer travel time with increasing offset/angle.

On a full-stack the dim spot can be clearly evident. If a clear reflection can be mapped on the data, then something like a dimming on the crest of a structure may suggest a DHI is present. For the dim spot to be trusted the interpreter must have a good understanding of the seismic acquisition and processing so that the amplitude behaviour can be attributed to change in fluid content rather than a simple lack of signal (i.e., decrease of the signal-to-noise ratio) at far offsets/angles. Looking for this type of response on full-stack data is certainly not recommended unless there is some strong well-calibration in the area. Vintage seismic or simple non-optimum data acquisition or processing can frequently lead to a localized dim area in the data. To distinguish

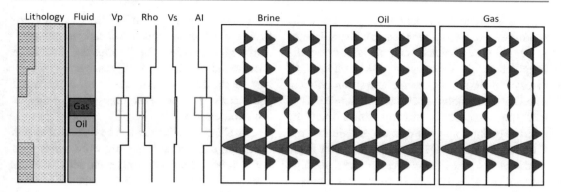

Fig. 6.7 Well based synthetic example using Gassmann subsitution of what the difference between a hydrocarbon Class 1 and a brine Class 1 response might look like

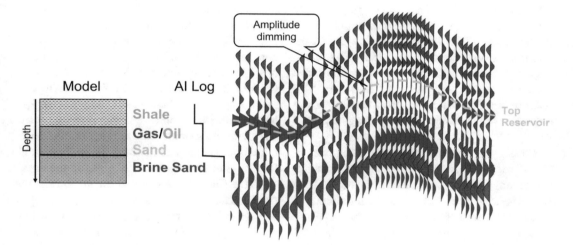

Fig. 6.8 Example of hard sand reflector (blue peak) becoming weaker at the top of a structural feature to indicate the presence of hydrocarbon shown by a class 1 AVO response

this as a hydrocarbon response, seismic gathers should ideally be present, so the interpreter looks for the hydrocarbon reflection amplitude to weaken more quickly and with greater magnitude than the brine-filled reservoir as evidence that the hydrocarbons are softening/reducing the impedance contrast. Figure 6.7 shows a well-based synthetics example of what a Class 1 response might look like against a brine case. Figure 6.8 illustrates the example of a dim spot in a stacked seismic section.

6.2.2 Class 2/2p—Bright Amplitude and Phase Reversal

This class of AVO behaviour is split into two slightly different responses: Class 2 and Class 2p. Class 2p is probably the most useful for relating variation in seismic amplitudes to a DHI because its response can be very anomalous compared with the surrounding seismic events. However, it is important to have access to seismic pre-stack gathers for validation. In the original work by Rutherford and Williams (1989),

Class 2 is included but not Class 2p (although its response is shown under Class 2). Class 2 tends to be mainly associated with oil field reservoirs simply because the effect on P-velocity is less extreme than for a corresponding gas column. The Class 2 response shows at near-angle offsets that the reservoir is slightly softer than the overlying shale and there is no amplitude contrast at zero angle offset. As the incident angle increases it causes the waves to travel through more of the reservoir layer and records a greater softening effect of the hydrocarbons. Low soft amplitudes at zero and near incident angles are observed, but with an increase in offset the strength of the response gets progressively stronger resulting in much brighter amplitudes at the far offsets. Class 2p differs by showing a hard interface response at zero and near offset which begins to weaken, or dim, with increasing offset distance but then displays what is termed as a polarity flip. The same event now becomes a soft response which continues to strengthen to a strong soft amplitude at the far angles.

Class 2p is very useful for the interpreter in identifying a DHI, however it comes with a number of important considerations. Firstly, during post-migration pre-stack gather conditioning a processing geophysicist may observe a phase reversal on the gathers as a sign of unflattened gathers and, as a consequence, overly flatten the gathers and remove the phase reversal. Secondly, the presence of multiple energy and high-angle noise can affect only the far angle amplitudes potentially giving the appearance of a Class 2p response. The final consideration when interpreting a Class 2p response is that the geoscientist must pick the far-angle soft reversed polarity event and not just the near-angle hard response, otherwise this will lead to inaccurate top reservoir surface-mapping and can have a knock-on effect on reservoir thickness calculations and volumetrics.

6.2.3 Class 3—Bright Spots

Class 3 is the type of the AVO behaviour that geoscientists are most familiar with and would expect to represent a DHI. Over the years,

geoscientists have used AVO Class 3 for identifying and de-risking many oil and gas prospects across global basins. On the flipside, because of past successes Class 3 AVO responses can mislead both the inexperienced and experienced geoscientist, because everything that is bright is not hydrocarbons.

For Class 3 AVO the sand is softer than the overlying shale with brine and even softer with hydrocarbon. So, at zero offset the seismic event may already look anomalous compared to other events in a gather or near-offset stack. With increasing offset/angle, the Class 3 becomes progressively brighter and the amplitude on the far-offset stack should be highly anomalous compared to other events, which typically will have weakened. One caution here is that zero and near-offset/angle traces are sometimes contaminated by high-amplitude noise and referred to as hot near traces. These high-amplitude traces can make a legitimate Class 3 AVO response appear less convincing when comparing to the expected amplitude behaviour. This is why careful post-migration gather conditioning is so important (Chap. 4) to remove issues like this in the data.

Often Class 3 anomalies show up so brightly on full-stack data that interpreters may simply neglect to inspect the individual partial stacks or pre-stack gathers for verification of this response. Frequently, in frontier exploration areas access to these pre-stack products is not possible. Certainly, many prospects may be defined using only a full-stack amplitude to attract potential interest for farm-in activities. Great caution is required here, and it may be better for the amplitude influence to be removed from any further interpretation or risking allowing other components of the prospect to be robustly evaluated.

The biggest risk of Class 3 AVO is the presence of low-saturation gas or 'fizz gas'. The presence of only a small (10–20%) amount of gas in the pore space of a rock can have large effects on reducing the V_P and density. This often occurs when gas has migrated through a formation and has left residual gas behind. The AVO response of 20% gas saturation is remarkably like that of 80%, with the latter clearly not being a commercial saturation for development. The tried and

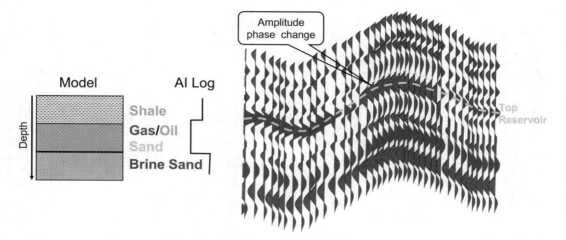

Fig. 6.9 Example of a hard sand reflector (blue peak) down-dip changing polarity to a soft sand reflector (red trough) at the top of a structural feature to indicate the presence of hydrocarbon shown by a class 2p AVO response

tested ways to manage and reduce uncertainties of drilling a Class 3 low-saturation gas prospect is as follows:

(a) Careful well-based AVO forward modelling;
(b) Use gas log data of previous low-saturation gas failures;
(c) Careful analysis of processing workflows applied to seismic data;
(d) Good understanding of migration pathways from petroleum systems analysis;
(e) Capture low saturation possibility in risking.

Figure 6.10 shows a P-velocity curve against water saturation with oil and gas cases in a sandstone and a similar plot for P-impedance.

Another pitfall that can manifest itself as a Class 3 AVO response is when the processing workflow applied to the data is not amplitude friendly. To deal with amplitude attenuation at depth caused by the absorption effects of the earth on the seismic wave, a common correction in a modern processing workflow is to apply a Q (quality factor, Chap. 4) correction to the amplitudes. This is far superior to the earlier days of applying AGC (Automatic Gain Control) and the dangers of short time windows producing false AVO. There can be dangers associated with applying Q-amplitude correction, particularly applicable for Class 3 AVO. Over correction of

amplitudes of an already soft event, such as a high-porosity brine sand, can lead to near and far events exhibiting a Class 3 AVO fluid response. It is good practice to apply any amplitude processing step as lightly as possible, and to always preserve a version of the pre-stack gathers and partial stacks without the correction applied for reference, and as a sense check.

Amplitude scaling with offset or angle in the gather domain should also be carried out carefully, particularly as it has a knock-on effect on any stack produced from the gathers. This tool can be very useful for recovering faded amplitudes. Some workflows will apply amplitude scaling in quite an aggressive way by using results from synthetic well-based forward AVO modelling to justify boosting seismic data to match these modelled responses. Changing the real data to fit a synthetic model is not advisable, but rather it is better to try to understand why the data and model do not agree—and leave it at that.

6.2.4 Class 4—Soft Event Getting Less Soft

Class 4 is the least understood and most difficult AVO behaviour from which to find true data examples of clear anomalies that have been

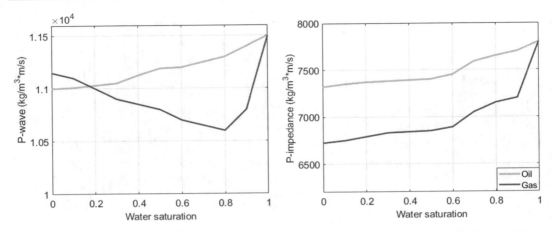

Fig. 6.10 Effect of the presence of hydrocarbon in (left) P-wave propagation velocity and (right) P-impedance

drilled and verified. Class 4 stands out from the other AVO classes by exhibiting a positive gradient, while the rest have negative gradients. A Class 4 reservoir is softer than the overlying shale, but with increasing offset the strength of the amplitude becomes weaker and can be almost absent at the far offsets/angles. These reservoirs have a strong, soft amplitudes at zero and near offsets/angles which are comparable or even stronger to that of a Class 3 response and are therefore likely to represent gas sands. So why do these sands give such a strong, soft response? Evidence shows that relatively shallow unconsolidated sands that are capped by much harder rocks, such as carbonates or heavily cemented sands and siltstones, can produce a Class 4 AVO anomaly. As the incident angle increases, the reflectivity response from unconsolidated sands becomes weaker and the shear component in the reservoir is also weaker than in a more competent reservoir. The same pitfalls relevant to Class 1 AVO also apply here. It is common for reflectors to lose amplitude strength with increasing offset/angle, especially at depth. However, a true Class 4 caused by gas should exhibit a much stronger soft amplitude response on the zero offset/angle than a brine-filled equivalent.

6.2.5 AVO Specific Seismic Conditioning

Once the geoscientist has identified a robust amplitude anomaly that exhibits a convincing AVO response, it is worth carrying out a seismic quality analysis or healthcheck before embarking upon more detailed and time-consuming analysis. The health check result may suggest that the seismic data would benefit from a conditioning workflow (Chap. 4), which should be implemented at this time. Additional volumes computed from seismic attribute analysis (e.g., quadrature trace) may be used to help refine the interpretation. Other post-migration seismic conditioning tools, like a structurally orientated filter or spectral blueing, may also be useful at this stage. Spending time producing a focused and detailed interpretation will ensure a cleaner seismic attribute map and add more technical rigour to the AVO anomaly.

6.3 DHIs, Flat Spots and Amplitude Conformance

The identification of particular types of AVO anomalies in seismic data can be enough to identify a particular area as worthy of further investigation. Screening large areas for AVO anomalies is a common technique in exploration and results can often be displayed on a single map, which helps with the first level of high grading for future focus areas. This method can also be used in development settings for near-field target screening, where the amplitude response is well understood through historically drilled wells and production. As the geoscientist focuses on the amplitudes in more detail there are several characteristics that transition an AVO anomaly from general interest into something far more compelling, which may help support the presence of a true DHI.

6.3.1 Flat Spots

Within a reservoir rock saturated by hydrocarbons the fluid is generally present at the shallowest depth (e.g., the crest of an anticline) where the reservoir meets the seal. As the volume of hydrocarbons entering the reservoir increases it builds downward from the shallowest point creating what is referred to as a column. The column of hydrocarbon does not always completely fill a trapping structure down to its base, which results in a contact where reservoir rock pores filled with hydrocarbon meet reservoir rock pores filled with formation water. A transition zone will be present where the hydrocarbon saturation slowly decreases over several metres before becoming absent.

The seismic reflection caused by the increase in acoustic impedance at the fluid contact can be very large, especially when going from a gas-saturated reservoir to a water-saturated reservoir. In some cases where both gas and oil are present two clear seismic events for each contact may be evident, however in practice this can be quite difficult because the gas column can have a very strong amplitude response which masks the gas-oil contact. In addition, the gas-oil contact will be a hard event so a sufficiently large enough oil column would be required to produce a soft response before getting another hard response, which represents going from oil to water. This

resultant seismic reflection event at a contact can appear horizontally flat as it obeys the law of fluid mechanics, where the less dense hydrocarbons float on the denser formation water. If the reservoir trap is structural in nature with dipping flanks, such as in an anticline 4-way dip, then the geometry of the contact and its associated seismic reflector appears very flat (Fig. 6.11). Flat spots are also seen in trapping examples that are not purely structural, where the reservoir may be a dipping layer which terminates against an updip fault seal. Cross-cutting reflectors (CCR) are similar to flat spots and follow the same principal. In this case the change in P-impedance caused by the different pore fluids properties will result in a strong, flat amplitude event seen to cut across other events such as the top and base of the reservoir layer.

Based on the mechanics of a flat spot you would imagine them to be commonplace, however high-quality flat spots on seismic data that represent the true position of hydrocarbon contacts are quite rare. Often, flat spots will simply be interpreted because the geoscientist sees an event that looks somewhat flat and it helps to validate a prospect and easily defines the base of the column. There are many other reasons why an interpreter might see an event that appears very flat:

(a) Erosional surface or unconformity.
(b) Seismic multiple (water bottom multiple).
(c) Cemented layer (secondary calcification).
(d) Hydrothermal mineralisation.
(e) Paleo contacts.
(f) Geology (flat reef).
(g) Out of plane reflector (two-dimensional seismic) or migration artefact.
(h) Velocity model layer artefact.
(i) Diagenetic alterations.

To fully validate the presence of a flat spot there should be integration between disciplines to get a thorough understanding of depositional environment, geological processes, petrophysical properties and seismic processing. These can help to mitigate the list of false flat spot pitfalls. Synthetic two-dimensional modelling can also be

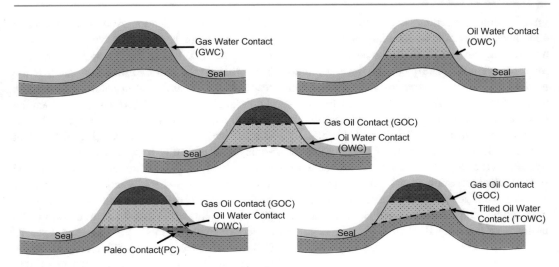

Fig. 6.11 A range of geo-models showing different hydrocarbon reservoir contacts

useful to understand whether a flat spot is possible and can allow the incorporation of rock properties from well data to make the model as accurate as possible.

6.3.2 Amplitude Conformance

During amplitude analysis the interpreter should make a range of seismic attribute maps in an attempt to understand the extent, magnitude, size and general characteristics of the amplitude anomaly. These attributes are mostly extracted at or around a mapped surface which is picked on the seismic data and represents a geological boundary in the subsurface (e.g., the top of a reservoir package or depositional event). This horizon will contain variation in elevation and topography which reflects components of the trap, whether structural of stratigraphic. How the amplitude anomaly, or more specifically the anomaly edge, conforms to the structure of the horizon has been proved to be one of the most critical observations as to whether an amplitude anomaly is really a DHI (Fig. 6.12).

Conformance is assessed by looking at the amplitude anomaly on a map with contours to indicate structural elevation. This amplitude is extracted from the seismic data using a horizon interpretation mapped on the reflector of interest.

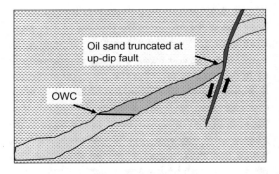

Fig. 6.12 Schematic representation of how a column of hydrocarbon fills down dip from the crest to a single conformance depth

For amplitude conformance analysis the interpretation should be made specifically focused on the anomaly event. Larger regional horizons are not fit for purpose at this stage of analysis. A regional surface may be interpreted on multiple seismic data sets or surveys, and is often either too coarse or, conversely, overly smoothed for the purpose of amplitude conformance analysis.

6.3.3 Time Versus Depth

A proportion of seismic interpretation in the industry is still carried out on seismic data in the time domain (TWT) which is then converted to depth using an appropriate velocity model. Looking at conformance in the time domain during the preliminary stages of conformance analysis is perfectly appropriate, but producing the same maps in the depth domain is an eventual necessity. The seismic and surface should be depth converted and the amplitude extraction performed, rather than depth converting a surface with an amplitude extracted in time. Structural features form traps, like 4-way dip closures, that flex and move when being converted from time to depth. The amplitude should ideally react with the structure and stay bounded by its edges. If during the depth conversion the structure appears to open out far beyond the amplitude leaving significant gaps at the trapping edges, or the amplitude is seen to spill out or cross-cut the edges of the structure, this is a clear concern. For amplitude prospects that have both structural and stratigraphic components or are just purely stratigraphic, the down dip amplitude conformance and, indeed, all edge conformance becomes even more important.

If we consider the case of a dipping reservoir-sand truncated by an up dip normal fault and sealed by shale above and below (Fig. 6.13), the lateral seal component might be either dip controlled or stratigraphically controlled (e.g., pinchout). From a down dip amplitude conformance perspective, we are looking for the amplitude to stop at a consistent contour which represents the contact and the extent of the hydrocarbons. Perfect or close to perfect amplitude conformance to structure is not uncommon where the trap has a structural component. Numerous real-world examples of this are seen for 4-way or 3-way dip closures against faults or salt diapirs.

When assessing amplitude conformance of a purely stratigraphically trapped prospect, for example a channel or turbidite sand lobe encased in a sealing lithology like a shale or mudstone, the interpreter should see more amplitude conformance variability simply because of the nature of system edges produced in these depositional environments, as shown in Fig. 6.14. It is also worth thinking about whether the seismic data has good enough resolution to image the pinched-out edges and the corresponding amplitude shut offs in this type of trap. Although it may be more challenging to get amplitude conformance for a purely stratigraphic trap, the interpreter should not loosen the rigour of the amplitude conformance analysis. Investing more time in gaining an integrated understanding of the geological depositional environment and possible spatial distribution of rock properties (e.g., porosity, permeability and sand quality) may help to explain the conformance that is observed. Geological analogs and supporting seismic attributes such as spectral decomposition and RGB blending are often essential for these anomalies to progress.

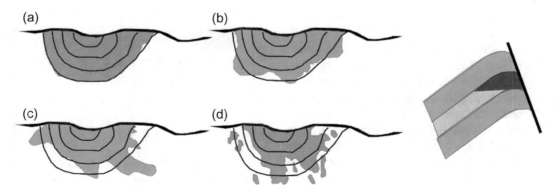

Fig. 6.13 Example of down dip amplitude conformance. **a** Excellent conformance along down dip extent at a constant depth contour. **b** Reasonable conformance to a constant contour with some variability due to seismic data quality. **c** Amplitude cross-cuts structure and is irregular in nature, some areas of limited conformance. **d** No visible down dip conformance and amplitude, cross-cuts structure and appears inconsistent

Fig. 6.14 Example of stratigraphic trap amplitude conformance. **a** Excellent conformance along all edges of the feature. **b** Reasonable conformance to a constant contour with some variability. **c** Amplitude cuts across several contours and is irregular in nature, some areas of limited conformance. **d** No visible edge conformance, amplitudes cross-cut structure and appear inconsistent

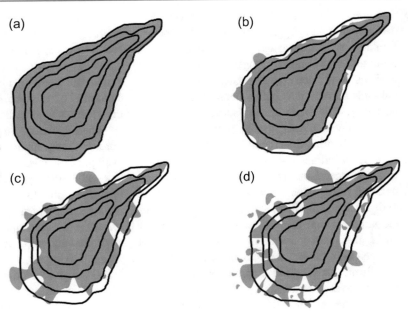

6.4 AVO Cross-Plot and Angle-Dependent Impedance

Section 6.2 introduces intercept and gradient and how their behaviours are used to describe the four AVO classes (Fig. 6.6). Although the strength of the seismic reflectivity is a key factor when assessing an amplitude anomaly, the geoscientist can also observe characteristics about the intercept and gradient values to help understand the anomaly in a more complete AVO attribute.

Castagna and Swan (1997) propose the intercept versus gradient cross-plot (or A vs B) method using the two-term Shuey approximation (Table 6.2). This method can use seismic postmigration gathers or partial angle stacks as input to create corresponding volumes of intercept and gradient. The intercept volume is not dissimilar from a near-stack reflectivity volume and on its own does not really give further insight into identifying AVO anomalies beyond the near-stack reflectivity. The gradient volume can however be very useful as a standalone volume in identifying areas in the seismic data where anomalous gradients exist. To produce a good gradient volume, the seismic events used,

whether gathers or angle stack, should be well aligned with as little moveout or multiple energy as possible. By cross-plotting the intercept and gradient volumes we can identify which class of AVO the data points belong to by using a template (Fig. 6.15). These data points can then be colour coded and highlighted back on the seismic data to either support an existing anomaly or as volume to search for gradient anomalies of potential interest.

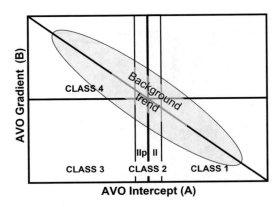

Fig. 6.15 Intercept (**a**) versus Gradient (**b**) cross-plot for AVO classes and background trend (adapted from Castagna and Swan 1997)

So far in this chapter we have focused on reflectivity strength viability with offset or angle to help identify AVO. Going beyond reflectivity a common progression in AVO analysis is to look at angle dependant impedance attributes, which are underpinned by intercept and gradient responses from the seismic data. Although seismic inversion (Chap. 7) is not the focus of this chapter, the topic of angle dependant impedance is relevant as it still ultimately relies on reflectivity changes with offset or angle, which is the basis of AVO.

Acoustic impedance volumes can be created using seismic inversion methods from full-stack or zero-offset seismic data. An acoustic impedance volume will give a very similar response to that seen in intercept volumes, however, true changes in amplitude strength relating to AVO are difficult to detect using acoustic impedance. Connolly (1999) showed how elastic impedance (EI) could be calculated for any angle of the incident (θ) (i.e., partial-stacks) and subsequently used to observe changes relating to AVO through inversion. The method is a deviation from a two-term linearization of the Zoeppritz equation that can predict EI over a pre-stack angle range of 0–30° following:

$$EI(\theta) = V_P^a V_S^2 \rho^c, \qquad (6.3)$$

where

$$a = k1 + \sin^2 \theta, \qquad (6.4)$$

$$b = 1 + \sin^2 \theta, \qquad (6.5)$$

$$c = \left(1 - 4k \sin^2 \theta\right), \qquad (6.6)$$

$$k = \left[\frac{V_s}{V_P}\right]^2. \qquad (6.7)$$

The EI method does have some limitations not seen with acoustic impedance. Elastic impedance values are technically dimensionless, varying in magnitude depending on the incident angle used, and therefore cannot be related to each other

across angle ranges or compared to measurable rock properties. Any errors in the velocity model used to generate the angle stack will result in the wrong elastic impedance being calculated.

The EI calculation was taken further by Whitcombe et al. (2002) with an approach that looked to provide a more quantitative elastic product which could be related to measurable petrophysical properties, such as those made by well-logs. This new method was termed extended elastic impedance EEI for the reason that it can be used for greater angles of incident than the previous EI calculation. It is also directly relatable to measurable petrophysical properties, allowing EEI logs to be calculated from well data. These logs act as calibration for EEI volumes generated from seismic data. EEI also differs from EI by using a rotation angle called chi (χ) rather than the more traditional incidence angle (θ) used in the EI method:

$$EEI(\chi) = [V_{P0}\rho_0] \left(\left(\frac{V_P}{V_s}\right)^{(\cos \chi + \sin \chi)} \left(\frac{V_s}{V_{S0}}\right)^{(8k \sin \chi)} \right.$$
$$\left. \left(\frac{\rho}{\rho_0}\right)^{(\cos \chi - 4k \sin \chi)} \right)$$

$$(6.8)$$

The EEI method is applied by using an acoustic impedance versus gradient impedance cross-plot generated from well data. The chi angle of rotation is optimized using well data in the AIGI cross-plot allowing different angles, or projections, to be calibrated for both lithology and fluid by matching them to well log responses i.e., V-shale, porosity, water saturation. Figure 6.16 shows the AIGI cross-plot with ortho-fluid projection from Whitcombe and Fletcher (2001).

Once a particular projection has been decided upon then the acoustic impedance and gradient impedance volumes can be generated from the seismic data and the interpretation can begin. This method is used extensively due to its ability to calibrate to petrophysical properties in wells, and also quickness of the workflow.

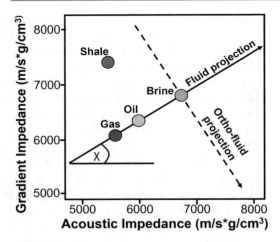

Fig. 6.16 AIGI cross-plot with rock properties data for different lithology and fluids showing projection angles (Whitcombe and Fletcher 2001)

6.5 Depth Trends, Compaction Trends and Overpressure Description

When looking at seismic amplitudes as indicators for the presence of hydrocarbons, it is common for a particular formation within an area of a basin to have had some historic success using seismic amplitude AVO anomalies. Although this is often the premise for many companies' exploration programmes and entries into new acreage, close attention needs to be paid by the interpreter to the similarity of the subsurface depths and pressure regime differences of any amplitudes they find compared to the discoveries using other amplitudes. The effect of reservoir cementing type, burial history and level of diagenesis that a reservoir rock has gone through over geological time will affect its P-wave velocity and consequently its AVO behaviour. This can lead to inaccurate comparisons of similar amplitude responses.

When looking for suitable hydrocarbon reservoir rocks, the physical properties of rocks in the subsurface vary depending on their current depth below the surface and more importantly their burial history. This is because the processes that occur immediately after deposition and

throughout burial can fundamentally change rock properties we see in the subsurface today. The same rock type or even the same formation can exhibit variations in rock properties over large basin scales but also over individual hydrocarbon field scales, sometimes down to variability within a single reservoir interval. These changes are controlled by, or a result of, the tectonic and sedimentary evolution of the target area.

The ability to identify and predict these physical rock properties and the presence of hydrocarbons through DHIs using seismic data will also vary depending on their depth. These variations in properties can be partially predicted by looking at depth trends. The geoscientist must be aware of how these may affect their ability to predict rock properties using seismic data. When using amplitudes from seismic data to either map geology or, further still, to make predictions about the quality and suitability of the rocks as reservoirs and the possible fluid fill, then depth trends should be well understood.

On a most basic level the geoscientist should be familiar with the concept that the deeper a rock is buried the more pressure or physical stress is exerted in all directions on that rock in the form of confining stress. This can be described following:

$$\rho g h, \tag{6.9}$$

where ρ is the density of the rock, g is the acceleration of gravity and h is the height of overburden above the object.

Using this over-simplified description of burial the general prediction can be made that the deeper the rock gets buried and the more pressure exerted, the tighter and more connected the framework of the rock becomes (Fig. 6.17). Therefore, properties such as velocity, density and P-impedance of rocks increase in general with depth as shown in Fig. 6.18, while properties such as porosity and permeability tend to decrease. This principle generally holds true and is a good starting point for understanding depth trends.

When looking to understand an amplitude response in one part of a basin compared to

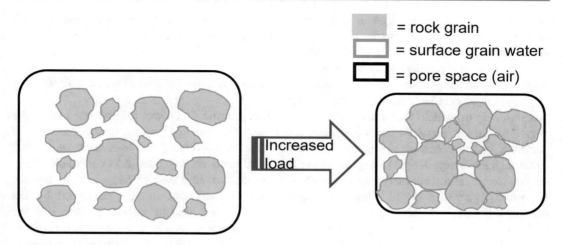

Fig. 6.17 The effect of applying load to a sediment particle and the increase in grain-to-grain contacts, reduction in pore space and increase water connectivity

Fig. 6.18 General depth trend plots for sediment undergoing burial

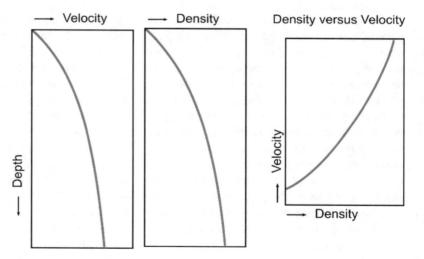

another, even for the same reservoir unit, it is crucial to understand the depositional and burial history and to rationalize the depth trends between the locations, also taking into consideration the current day depth variability. This is true for both exploration and more developed areas with some limited well control in appraisal and development stages. There are several real-world examples of discoveries driven by amplitudes or AVO at a particular depth. However, successive drilling of deeper or shallower amplitude anomalies turned out to be water wet.

This clearly highlights one of the major pitfalls of amplitude interpretation: the response is not unique, is uncertain and can vary with depth.

6.5.1   Deposition and Diagenesis

Generally, to understand depth trends for suitable hydrocarbon reservoir rocks we are mostly concerned with the behaviour of sedimentary rocks. The depositional system, processes and type of geological material (i.e., mineral composition) deposited in a sedimentary system will have a large initial control on key factors such as grain size, grain contacts, grain coatings, sorting, porosity and permeability. Having sound geological understanding of the source of sedimentary material is vital as it represents the starting properties from which the sedimentary rocks vary. The surface and near-surface conditions for pressures and temperature are generally much lower than those experienced during burial in the subsurface. So, at surface we expect sediments to be loosely packed with high-porosity, high-permeability and high-water content.

Diagenesis is the process that describes physical and chemical changes in rocks due to burial, occurring at fairly low pressures and temperatures (0–200 °C, 0–10 km). As the sediments are deposited and burial begins, the process of diagenesis becomes important to understand how physical and chemical changes affect rock properties, which, in turn, affect seismic responses and suitability for detecting the presence of hydrocarbons accurately within the pore space using AVO.

The effects and rate of diagenesis within a sedimentary basin vary greatly depending on factors such as sediment supply into the basin, topography and active structural tectonics during deposition; all of which may facilitate either slow or fast burial speeds. There are three main processes that influence the properties of sedimentary rocks generated from deposited sediments:

(a) Early post-deposition and consolidation;
(b) Compaction;
(c) Lithification.

Figure 6.19 shows how the three stages listed above can be represented within the subsurface and the effect on the grains and pore spaces with increased burial leading to increased temperature and pressure.

When new mineral crystals grow in the pore space of sediments the process is called cementation. This process lithifies the sediments into rock and leads to a reduction of porosity and permeability, which are two of the most important properties of reservoir rocks. The type and extent of cement can be paramount for predicting

rock properties and also in understanding AVO responses in potential reservoir rocks.

6.5.2 Overpressure Effect on AVO

The pressure that rocks are subjected to in the subsurface can play an important role when it comes to identifying elastic rock properties and AVO using seismic data. AVO anomalies identify changes in a rock's compressional velocity, shear velocity and density influenced by the lithology and fluid properties. Variations in the pressure regime can lead to the seismic response behaviour appearing quite different from what is expected, resulting in a more uncertain interpretation of AVO anomalies.

Lithostatic pressure, also referred to as overburden pressure, describes the pressure put on a rock layer by the weight of the overlying overburden. Pressure is transferred through grain-to-grain contact through the successive rock layers. The amount of pressure is determined by the density and thickness of the overlying geology. Hydrostatic pressure describes the pore fluid pressure within a rock's pore space. This pressure is calculated down from sea or ground levels and gives the expected pore pressure for a given depth of the rock, which can be calculated as a linear trend. In a fully saturated rock, the pore space is completely full of fluid. The fluid within the pores exerts a uniform pressure in all directions on to the grains of the matrix of the rock (i.e., the pore pressure). This pressure acts as a supporting resistance force for the rock matrix pore space against the confining lithostatic pressure.

If a stratigraphic layer becomes isolated or disconnected, meaning the pores are no longer in communication with the larger rock layer, the pore water pressure characteristics can become altered and are no longer in hydrostatic equilibrium. The result is rocks becoming under or overpressured compared with the surrounding strata. This process can occur in both non-reservoir lithologies like shales and reservoir lithologies such as sands. Figure 6.20 illustrates the difference in hydrostatic and lithostatic pressure.

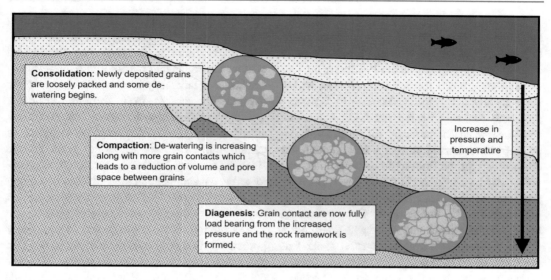

Fig. 6.19 Main stages of burial that change the deposited sediment into a sedimentary rock (adapted from Fan et al. 2019)

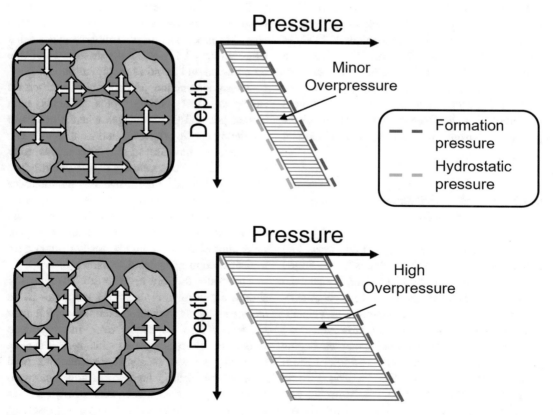

Fig. 6.20 (Top image) The difference that minor over-pressure makes may be present when a formation is close to equilibrium with hydrostatic pressure. (Bottom image) The pore fluid pressure is much greater which causes the formation pressure to be much higher than the standard hydrostatic pressure at that depth (adapted from Goffey et al. 2020)

If a reservoir rock layer becomes overpressured the pore pressure increases which can lead to a reversal of the effect of the lithostatic pressure. This reversal reduces the P-velocity for that isolated area, while the seismic response would see a localized P-velocity decrease in the overpressured rocks. A search for AVO anomalies would show an anomalously soft response compared to the surrounding rocks, even though the effect might not be from hydrocarbon fluids, and potentially be a false AVO signal. Overpressured shale can also lead to AVO anomalies and elastic properties that do not relate to the presence of hydrocarbons. Ramos and Toledo (2004) show a drilled example of an overpressured shale that has much lower P-impedance and Poisson's ratio compared to sands at the same depth. The overpressured shale also gives a large contrast when overlying a high P-impedance sand, which provides a strong AVO gradient indicative of a hydrocarbon response. Cap-rock shales tend to hold water during burial and so are more likely to be affected by pore pressure, it is therefore important to model this parameter when modelling rock properties. It is certainly possible for identical reservoir sands to exhibit very different amplitude and AVO entirely due to the pore-pressure environment (Lindsay and Koughnet 2001).

6.5.3 Porosity Preservation in Deeply Buried Rocks

Deeply buried sandstones generally have low porosities based on the mechanisms described previously in this chapter. Therefore, because the stress of burial reduces porosity we expect the P-velocity of the rock to be increased as velocity changes inversely with porosity. However, there are many examples of sandstones at depths greater than 2–3 kms that exhibit helium porosities in a range of 25–35%. The Triassic Skagerrak formation from the UK North Sea and the Upper Cretaceous sandstone of the Santos Basin offshore Brazil both exemplify this finding.

The presence of clay during deposition can coat quartz grains and restrict quartz overgrowth, although factors such as temperature during burial may vary the effectiveness for the limitation of this growth. Research has shown that coating the sandstone matrix grains with microcrystalline quartz may also inhibit larger quartz overgrowth cements from forming. Sufficient quantities of certain sedimentary materials such as siliceous microfossils may be required in certain concentrations to make available the siliceous material for this microcrystalline quartz to grow on sediment grains, ultimately linking the ability for this kind of porosity preservation to specific sedimentological environments.

Dissolution and the generation of secondary porosity is another proven method not to be ruled out when observing high porosities at great depth. The dissolution of minerals such as feldspars can be confirmed by looking at thin sections and scanning electronic microscopy (SEM) of reservoir rocks, as well as comparing modelled porosities with what is found in well core. The process of this dissolution suggests that a system must be open for the dissolved fluids to be removed and avoid re-precipitation.

This phenomenon of high-porosity reservoir sands at great burial depths presents both advantages and disadvantages when it comes to predicting elastic properties and using AVO. The fact that an increase in porosity relates to a reduction in P-wave velocity means that a brine-filled high-porosity sand, caused by overpressure or porosity preservation mechanism, will give a strong, soft amplitude response in seismic data and a false AVO signal. This is likely to appear quite anomalous and could be easily mistaken for a positive sign of hydrocarbons.

However, the preservation of the porosity should allow for the pore fluid properties to be expressed in the seismic response, which may allow AVO to be useful even in very deep reservoirs. In recent years this type of deep reservoir, known as high pressure and high temperature (HPHT), has been of great interest to large energy companies. Historically it was thought that at these depths only poor reservoir properties existed and, so, combined with the high temperatures, high pressures, challenging drilling conditions and large economic cost there was little value in exploration. However, these conditions are perfect for a type of hydrocarbon reservoir which is able to hold huge columns of hydrocarbon, because of the presence of over-pressured shale seals whose capillary entry pressure is very difficult to overcome.

6.6 Practical Guide to Seismic AVO Analysis

The reliability of the AVO interpretation depends mainly on two characteristics. The experience of the interpreter geophysicist and the quality of the seismic data. The former influences the quality of amplitude maps, as good quality seismic inter-pretation will help to give the most accurate mapping of amplitude anomalies The latter depends on the survey acquisition and processing workflow applied to the data (e.g., Q-amplitude compensation). These are important for validating amplitude anomalies against amplitude artefacts. Gather and partial stacks will benefit from having focused AVO conditioning workflow applied which will increase confidence in any amplitude anomalies. To validate observed AVO anomalies, forward modelling using well data should be used. Finally, depth trends and overpressure can give misleading amplitude responses and should be fully understood within an area of interest.

References

Aki K, Richards PG (1980) Quantitative seismology: theory and methods. W. H.. Freeman and Co

Bortfeld R (1961) Approximations to the reflection and transmission coefficients of plane longitudinal and transverse waves. Geophys Prospect 9(4):485–502

Bunt RJW (2015) The use of seismic attributes for fan and reservoir definition in the Sea Lion Field, North Falkland Basin. Petrol Geosci 21(137–149):23. https://doi.org/10.1144/petgeo2014-055

Castagna JP, Swan HW (1997) Principles of AVO cross plotting. Lead Edge 16(4):337–344. https://doi.org/10.1190/1.1437626

Chopra S, Marfurt KJ (2005) Seismic attributes—a historical perspective. Geophysics 70: 3SO–28SO

Churlin VV, Sergeyev LA (1963) Application of seismic surveying to recognition of productive part of gas-oil strata. Geologiya Nefti I Gaza 7(11):363

Clay CS, McNeil H (1955) An amplitude study on a seismic model. Geophysics 20:766–773

Connolly P (1999) Elastic impedance. Lead Edge 18:438–452

Dobrin MB (1976) Introduction to geophysical prospecting, 3rd edn. McGraw-Hill

Fan F, He L, Zhou B, Song X (2019) Hydrothermal-assisted transient binder jetting of ceramics for achieving high green density. JOM 72(3):1307–1313

Fatti JL, Smith GC, Vail PJ, Strauss PJ, Levitt PR (1994) Detection of gas in sand reservoirs using AVO analysis: a 3-D seismic case history using the Geostack technique. Geophysics 59:1362–1376

Goffey G, Gluyas J, Schofield N (2020) UK oil and gas fields: an overview. Geol Soc Lond Memoirs 52(1):3–18

Hilterman FJ (1975) Amplitudes of seismic waves—a quick look. Geophysics 40:745–762

Hilterman FJ (1983) Seismic lithology: presented as a continuing education course at the 53rd Ann. Society of Exploration Geophysicists, Tulsa, OK, Internat. Mtg.

Hilterman FJ (2001) Seismic amplitude interpretation. Society of Exploration Geophysicists, Tulsa, OK

Klarner S, Ujetz B, Fontana RL (2008) Enhanced depositional and AVO models for lithologically complex sandstones in the Santos Basin, offshore Brazil. Pet Geosci 14:235–243. https://doi.org/10.1144/1354-079308-760

Levin FK, Bayhi JF, Dunkin JW, Lea JD, Moore DB, Warren RK, Webster GM (1976) Developments in exploration geophysics, 1969–74. Geophysics 41:209–218

Lindsay R, Koughnet RV (2001) Sequential backus averaging: upscaling well logs to seismic wavelengths. Lead Edge 20(2):188–191. https://doi.org/10.1190/1.1438908

Mann RL, Fatt I (1960) Effect of pore fluids on the elastic properties of sandstone. Geophysics 25(2):433–444. https://doi.org/10.1190/1.1438713

Ostrander WJ (1984) Plane wave reflection coefficients for gas sands at nonnormal angles of incidence. SEG Technical Program Expanded Abstracts, pp 216–218

Paulson KV, Merdler SC (1968) Automatic seismic reflection picking. Geophysics 33:431–440

Phillips TB, Jackson CA-L, Bell RE (2019) Rivers, reefs and deltas: geomorphological evolution of the Jurassic of the Farsund Basin, offshore southern Norway. Pet Geosci 26:81–100

Ramos ACB, Toledo MAS (2004) Induced AVO anomalies from pore pressure effects. SEG Technical Program Expanded Abstracts, pp 252–255

Rutherford SR, Williams RH (1989) Amplitude-versus-offset variations in gas sands. Geophysics 54:680–688. https://doi.org/10.1190/1.1442696

Shuey RT (1985) A simplification of the Zoeppritz equations. Geophysics 50:609–614

Smith GC, Gidlow P (2003) The fluid factor angle and the crossplot angle. 73rd Annual International Mgg., Society of Exploration Geophysicists, pp 185–188

Whitcombe DN, Connolly PA, Reagan RL, Redshaw TC (2002) Extended elastic impedance for fluid and lithology prediction. Geophysics 67:63–67

Whitcombe DN, Fletcher JG (2001) The AIGI crossplot as an aid to AVO analysis and calibration. SEG Technical Program Expanded Abstracts, pp 219–222

Wyllie MRJ, Gregory AR, Gardner LW (1956) Elastic wave velocities in heterogeneous and porous media. Geophysics 21:41–70. https://doi.org/10.1190/1.1438217

Yilmaz O (1987) Seismic data processing. Society of Exploration Geophysicists, Tulsa

Zoeppritz K (1919) Erdebebenwellen VIIB. Über Reflexion and Durchgang seismischer Wellen durch Unstetigkeitsflächen. Nachrichten Von Der Gesellschaft Der Wissenschaften Zu Göttingen, Mathematisch-Physikalische Klasse I:66–84

Predicting Subsurface Properties from Seismic Data

7

Seismic reflection data play a central role in the subsurface modelling and characterization steps of the geo-modelling workflow, as they provide exhaustive spatial information about the subsurface geology. The recorded seismic amplitudes are indirect measurements of the subsurface rock and elastic properties (e.g., porosity, volume of mineral, fluid saturations, density, P-wave and S-wave velocities) and predicting these properties from seismic amplitudes involves solving a challenging inverse problem (i.e., seismic inversion) (Tarantola 2005).

Inverse problems have a huge range of applications including science, mathematics, engineering, computer and finance to name but a few. It is a vast and complex area, which is still a hot research topic in earth sciences. The application of inverse theory has been used for many years within the field of applied geophysics especially within the energy industry. When approaching an inverse problem, we simply take a set of indirect observations, or measurements, and try to predict the cause (i.e., the model parameters) of these measurements. Inverse problems help us to understand parameters and conditions that we cannot directly measure. As mentioned in the previous chapters seismic data are not a direct measurement of the subsurface, we can use inverse theory to model the subsurface quantitatively.

Scales et al. (2001) give a very simple but clear example of the application of inversion (Fig. 7.1). The concept is based on making several measurements at the surface with the purpose of finding something we believe to be present in the subsurface, but for which the spatial location is unknown. The measurement we are able to make (e.g., gravity) does not directly relate to what we are searching for (e.g., treasure), however, the presence of the object we are looking for influences the physical property we can measure.

In Fig. 7.1, we are looking to find a relationship between gravity and another physical property that we favour more as a direct indicator of what we are looking for (i.e., density). Having a density map of the island would help to find the treasure. The challenge is to calculate numerically a number of synthetic density models of the subsurface to see which of these would produce a gravity profile to match the field observations. From a simplistic perspective, to do this we need to build a three-dimensional cellular grid model that represents the subsurface and then populate each cell with discrete values.

The first issue we will encounter when coming up with the correct model is that of non-uniqueness. This means that multiple synthetic subsurface models can be produced which will give a similar match to the field data observation. This is a clear problem and one of the main reasons why using a deterministic inversion approach producing a single modelled result is often less favoured compared to statistical-based (or stochastic) inversion methods that capture uncertainty through multiple modelled results.

© The Author(s), under exclusive license to Springer Nature Switzerland AG 2022
T. Tylor-Jones and L. Azevedo, *A Practical Guide to Seismic Reservoir Characterization*,
Advances in Oil and Gas Exploration & Production, https://doi.org/10.1007/978-3-030-99854-7_7

Fig. 7.1 Shows a gravity measurement location and profile for buried gold under the sand (adapted from Scales et al. 2001)

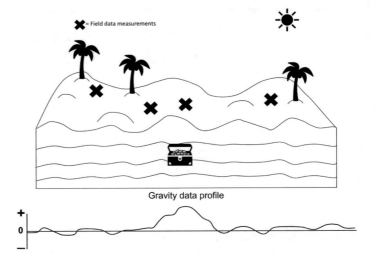

However, this non-uniqueness can be tackled in a number of ways. Firstly, some models produced during an inversion may be so far from what is physically possible, or likely, that they can be discarded using simple criteria-based constraints within the inversion. There are two general groups of constraints that are often applied: hard constraints and soft constraints.

Soft constraints guide the inversion model building and can be broken if certain criteria are met. These softer constraints are often concerned with keeping the model within the laws of known physics and controlling large erratic changes in the magnitude of the estimated property over small spatial distances and depths. Rock property relationships can be constrained in this way and it is common in seismic inversion to use what is termed a mudrock line (Castagna et al. 1985), or equivalent empirical linear relationship between P-wave and S-wave velocity, for brine saturated siliciclastic rocks. The Gardner line (Gardner et al. 1974) is another commonly used empirical relationship between density and velocity based on laboratory measurements for brine-filled rocks.

Hard constraints impose a much greater influence on an inversion model and although discouraged by some practitioners the use of hard constraints can be necessary. When applied carefully and methodically they may help an inversion model to continue down the

metaphorical inversion space fairway, rather than repeatedly getting stuck in local minima. Hard constraints often come in the form of geological boundaries, layers or geo-bodies that are inserted into either a starting model or intermediate iteration model, which has not converged sufficiently. These inserted features can be given specific values based on prior knowledge of the subsurface. Hard constraints do however come with a dilemma because it could be argued that they can bias the result of the inversion from being purely driven by the input data.

Another way to validate models is to use prior information. Prior information is an independent observation to the primary input data being inverted and can come in the form of many different inputs. Within the energy industry the best prior information is obtained from well data, which represent direct measurements of the subsurface. Seismic data can also be used as spatial prior information, if trusted, for some geostatistical methods using co-kriging techniques between well and seismic data. The prior comes with the same dilemma as the hard constraints, because the user influences the inversion process. The geoscientist may have a model in mind and this bias can be unconsciously imposed on the inversion when applying prior information.

The most important consideration for any inversion is the main input, which is the field data, or in the case of this book the seismic data. If the data being inverted is not able to detect, measure or resolve the feature(s) of interest then the inversion cannot make these geological features simply appear. All too often the inversion processes are proposed to solve a problem for which the input data are unsuitable. In the energy industry this has led to polarizing views of inversion techniques as tools to bring value. To solve this problem either the input data need to be improved for a single deterministic result or additional input data will need to be added and used to generate multiple results capturing spatial uncertainty in the probabilistic approach.

This chapter introduces the mathematical background regarding seismic inversion followed by a description of the current trend in seismic inversion methods, with a focus on geostatistical methods, and best practices in reservoir geophysics.

7.1 Seismic Inversion

In seismic inversion one aims at predicting the spatial distribution of continuous ($\mathbf{m}$) (e.g., acoustic and elastic impedances, P- and S-wave velocities and density) and/or categorical subsurface properties ($\mathbf{f}$) (e.g., facies and rock types) from recorded seismic amplitudes $\mathbf{d_{obs}}$. Within this context we often define facies as the combination of the solid and fluid phases of the rocks and not strict sedimentary facies. These are often designated as lithofluid facies (as defined by Avseth et al. 2005) in the geophysical literature. The forward model, which maps the model parameters into the data domain can be summarized by:

$$\mathbf{d_{obs}} = \mathbf{g}(\mathbf{m}, \mathbf{f}) + \mathbf{e}, \qquad (7.1)$$

where $\mathbf{g}$ is the known forward model, which describes the subsurface propagation of the acoustic wavefield generated by the seismic source, and $\mathbf{e}$ represents noise in the data and physical approximations assumed during data processing (Tarantola 2005).

On the other hand, seismic inversion (i.e., obtaining the inverse function $\mathbf{g}^{-1}$ that allows to retrieve $\mathbf{m}$ and $\mathbf{f}$ from $\mathbf{d_{obs}}$) is unknown, nonlinear, ill-posed and allows for multiple solutions (Tarantola 2005). This means that there are multiple combinations of models $\mathbf{m}$ that generate similar synthetic seismic data to the observed data ($\mathbf{d_{obs}}$). In practical terms, ensuring a good match between synthetic and observed seismic data is not enough to ensure the geological plausibility of the inverted models. The predicted models might match the observed data but be considerably different from the true subsurface geology.

Depending on assumptions about the model parameters and the forward model ($\mathbf{g}$), the existing seismic inversion methods may be classified into two classes: deterministic and stochastic (or statistical-based) methods. A detailed mathematical description of these seismic inversion methods can be found in detail in Doyen (2007), Bosch et al. (2010), Azevedo and Soares (2017) and Grana et al. (2021).

In this book we focus on stochastic seismic inversion methodologies, and specifically on iterative geostatistical seismic inversion methodologies, for two main reasons. First, these inversion methodologies have proven their value to predict the spatial distribution of the subsurface rock and elastic properties below the seismic resolution and their applicability has grown within the energy industry in the last decades. Second, the understanding of these methods is still limited and we aim to contribute to the generalization of these methods.

Deterministic seismic inversion methods (e.g., Oldenburg et al. 1983; Russell 1988; Russell and Hampson 1991) are still the seismic inversion methods most applied in the industry as they are computationally cheaper, simpler and easier to parameterize than the statistical-based seismic inversion methodologies. Additionally, the energy industry has an in-depth understanding of these inversion methods as they have been successfully used for more than thirty years.

In general, these methods start from an initial model, commonly a low-frequency model

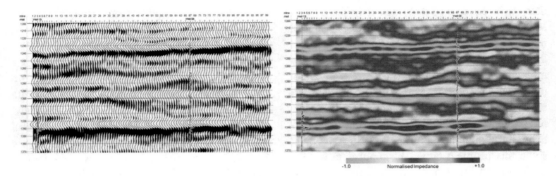

Fig. 7.2 Conventional seismic reflection data (Left) with well-logs compared to coloured inversion impedance (Right). The sand intervals are shown as the blue/purples (Francis 2013). (Courtesy of GEO ExPro)

(Sect. 7.2) and use deterministic optimization algorithms to update the initial model iteratively while converging the predicted seismic reflection data towards the observed one. Different deterministic seismic inversion methods have alternative implementation details and are grounded in different assumptions (Sect. 7.3). Nevertheless, the predicted subsurface models from deterministic inversion methodologies are always smooth representations of the subsurface (i.e., thin geological layers below the seismic resolution are not properly detected), representing the average behaviour of the subsurface properties at the seismic scale. Additionally, the solution provided by these methods has limited uncertainty assessment capabilities.

Statistical-based inversion methods are normally divided into Bayesian linearized approaches (e.g., Buland and Omre 2003; Grana and Della Rossa 2010; Grana et al. 2017) (Sect. 7.4) and those relying on stochastic optimization methods such as genetic algorithms, geostatistical simulation and co-simulation (Sect. 7.5), simulated annealing and neighbour algorithm (e.g., Sen and Stoffa 1991; Bortoli et al. 1993; Sambridge 1999; Le Ravalec-Dupin and Noetinger 2002; Soares et al. 2007; González et al. 2008; Grana et al. 2012; Azevedo et al. 2015, 2018; Azevedo and Soares 2017; Grana et al. 2021). Stochastic optimization methods allow sampling the model parameter space directly producing alternative realizations of the subsurface properties of interest. These methods result

in an ensemble of models that represents the spatial uncertainty about the predictions.

As each seismic inversion methodology has its own assumptions, limitations and advantages, the selection of the most suitable seismic inversion method should be site-dependent and based on the (Sect. 7.6):

(a) geological complexity of the study area;
(b) number of available wells and the existing well-log suite;
(c) quality of the seismic reflection data.

7.2   Low-Frequency Model Building

Due to the limited nature of the recorded seismic bandwidth, and despite the recent advances in seismic acquisition and processing techniques on broadband seismic (Sect. 3.4.5), the signal associated with the low-frequency band, below approximately 5 Hz, is either not present in the observed seismic data or have low signal-to-noise ratio. Therefore, this information is hard to recover directly from the observed seismic inversion. For this reason, most of the seismic inversion methods available in commercial software packages need to explicitly include a low-frequency model that consists of the expected background vertical and horizontal trends of the subsurface properties to be inverted.

From a conceptual perspective, the low-frequency model is a representation of the a priori geological knowledge about the subsurface and helps constraining the space of plausible geological models (i.e., it acts as a spatial regularizer of the inverse solution). This information (i.e., the full bandwidth of the seismic data) is needed to correctly predict the absolute rock and elastic properties of the subsurface. The lack of full bandwidth results in the non-unique nature of the seismic inversion problem: there are multiple Earth models that generate synthetic data similar to the observed one, some of these models may represent unplausible geological models given the known geological setting. The integration of the background model tries to avoid this situation.

A general workflow to generate these low-frequency models might comprise the following steps. Filtering the existing well-log data for the properties of interest at the desired band of frequencies. The filtered logs are then interpolated and extrapolated extended for the entire inversion grid. The interpolation of the well-log data might use different interpolation algorithms (e.g., geostatistical interpolation, inverse distance weighting and nearest neighbour) and follow a given stratigraphy as interpreted from the seismic reflections and/or be conditioned to velocity models resulting from conventional velocity analysis on CMP gathers (Yilmaz 1987) and/or include rock physics depth trends (Avseth et al. 2005). Geostatistical interpolation techniques such as kriging (Deutsch and Journel 1992) might be preferable tools to perform this task as they ensure the reproduction of the well data at the well locations, allow to integrate information regarding the expected spatial continuity pattern using a variogram model and to integrate secondary information such as velocity models from seismic velocity analysis. In seismic inversion directly for rock properties (Chap. 8) these models should be built conditioned to rock physics depth trends (Avseth et al. 2005; Martinho et al. 2021). In cases where no well data are available, the low-frequency models might be the ones resulting from conventional seismic velocity analysis.

7.3 Deterministic Seismic Inversion

There are several deterministic seismic inversion methods implemented in commercial geo-modelling software widely applied in the energy industry with differences in their implementation and in the assumptions behind each implementation. In this section we do not intend to provide a detailed overview of deterministic seismic inversion but provide an overview of the more general deterministic inversion methods.

The simplest deterministic seismic inversion methods to consider are recursive (e.g., Russel 1988) and coloured inversion (Lancaster and Whitcombe 2000). These methods can be considered as 'fast-track' approaches to predict acoustic impedance from seismic data. Though considered as inversion methods in the energy industry they do not solve the seismic inversion problem as previously discussed, rather they represent an approximation to this problem. Both methods mentioned herein can be useful for a preliminary model about the spatial distribution of the acoustic impedance in early stages of the geo-modelling workflow.

Recursive inversion is the oldest seismic inversion method and tries to predict absolute acoustic impedance values from fullstack seismic reflection data. However, it does not consider the effects of the wavelet, and attempts to scale the seismic reflection to reflectivity based on an initial low-frequency model following:

$$I_{P_{i+1}} = I_{P_i} \frac{1 + r_i}{1 - r_i}.$$

This equation represents the inverse of the normal incidence reflection coefficient.

Coloured inversion is based on well and seismic reflection data. In this simple approach one operator is constructed between the amplitude spectrum of the recorded seismic with the relative acoustic impedance as retrieved from the well information. This operator is then convolved with the recorded seismic to result in a predicted acoustic impedance model. The biggest assumption of this method is that the amplitude spectrum of the recorded seismic is stationary, which is not valid for large volumes and inversion time gates. Besides, coloured inversion assumes the sparse information retrieved from the well data is also stationary and valid for the entire seismic volume (Fig. 7.2).

Although the coloured inversion cannot be directly related to petrophysical properties such as lithology or fluid type it does have several advantages over the more complex inversion methodologies, which make it a useful analysis tool for the geoscientist. The workflow is quick and requires no specialist geophysical knowledge. Coloured inversion requires only basic well-logs to design and create an operator that is convolved with the data. Well ties and wavelet estimation used to understand whether the seismic data are zero phase are not required for coloured inversion and the results are, therefore, less influenced by wavelet issues. Coloured inversion results are consistent with more advanced deterministic inversion results (sparse-spike constrained and unconstrained), which require extracted wavelets and a low-frequency model component.

Model-based and sparse-spike approximations of deterministic seismic inversion methods (e.g., Oldenburg et al. 1983; Russell 1988; Russell and Hampson 1991) have been the preferred methods applied in the energy industry for subsurface modelling and characterization. These inversion methods have been developed to work in both the acoustic and elastic domains considering post- and pre-stack seismic data. These deterministic inversion methods are trace-by-trace approaches and use deterministic optimization techniques, such as gradient descent, to minimize the error between observed and predicted data. The inverted model (**m**) can be found by minimizing an objective function $\mathbf{J}(\mathbf{m})$ representing the norm $\|.\|$ (e.g., L1- and L2-norms):

$$(\mathbf{m}) = \text{argmin}_{\mathbf{m}}\mathbf{J}(\mathbf{m}) = \text{argmin}_{\mathbf{m}}\|\mathbf{d} - \mathbf{g}(\mathbf{m})\|^2. \tag{7.2}$$

In model-based seismic inversion, an initial subsurface model (Sect. 7.2) is perturbed

iteratively until the synthetic seismic calculated after each perturbation is close enough to the observed one (i.e., the difference between synthetic and recorded data is below a pre-defined threshold). The objective function is normally defined as the L2-norm between observed and synthetic traces, which is modified to include a regularization term to ensure lateral continuity and avoid sharp discontinuities between adjacent traces and to keep the space of solution within the range of geologically plausible models (Russell 1988; Russell and Hampson 1991).

The sparse-spike approximation is based on the deconvolution of the seismic trace under a sparse assumption and assumes that the observed seismic trace can be matched with a minimum number of reflection coefficients. First, a series of sparse reflection coefficients are predicted from the seismic data by deconvolving the seismic trace with a known wavelet. These reflection coefficients are constrained to the initial model and used to compute the elastic property of interest. The Maximum Likelihood (Hampson and Russell 1985) and the L1-norm (Oldenburg et al. 1983) methods are two of the most well-known examples of this family of deterministic seismic inversion methods.

In general, the model update step in deterministic seismic inversion methodologies uses nonlinear minimization approaches such as gradient-based, which linearize the inverse problem around the initial solution as described by the low-frequency model. Beside the widespread use of these methodologies within the industry, the uncertainty assessment of deterministic solutions is limited as it can only be assessed by a linearization around the best-fit inverse solution (i.e., under Gaussian assumption). Also, the inverted subsurface models are smooth representations of the true subsurface geology, do not ensure the reproduction of the well-log data at the well locations and represent the average behaviour of the subsurface. By predicting a smooth version of the true geology these methods do not reproduce extreme scenarios (e.g., high and low porosity and permeability regions, which are critical for fluid flow) that are required for risk analysis. Figure 7.3 shows an illustrative example of a trace-by-trace deterministic seismic inversion method, which allows the simultaneous prediction of (I_P) and P-wave and S-wave velocity ratio (V_P/V_S) from partial angle stacked volumes. This simultaneous inversion method is based on the sparse-spike approximation (Russell 1988) where a nonlinear optimizer is used at each trace location sequentially. The perturbations are performed from initial low-frequency models of the elastic properties of interest (Fig. 7.3a). Figure 7.4 compares a deterministic inversion model against three geostatistical realizations. The deterministic model is smoother and unable to capture the small-scale variability of the stochastic inversion models.

7.4 Bayesian Linearized Seismic Inversion

In Bayesian linearized seismic inversion the posterior distribution of the model parameters can be computed analytically under some statistical assumptions about the prior and errors present in the data and assuming a linear forward model (e.g., Buland and Omre 2003; Grana and Della Rossa 2010). In this case, the posterior probability distribution can be computed following:

$$P(\mathbf{m}|\mathbf{d}) = \frac{P(\mathbf{d}|\mathbf{m})P(\mathbf{m})}{P(\mathbf{d})}, \qquad (7.3)$$

where $P(\mathbf{m})$ is the prior distribution of the model $\mathbf{m}$, $P(\mathbf{d}|\mathbf{m})$ is the likelihood function, and $P(\mathbf{d})$ is a normalizing constant to ensure that is a valid probability density function (i.e., $\int P(\mathbf{m}|\mathbf{d}) = 1$).

If we assume a linear forward model, a Gaussian prior distribution of the model parameters $\mathbf{m}$, $\mathbf{m} \sim \mathcal{N}(\mu_\mathbf{m}, \Sigma_\mathbf{m})$, where $\mu_\mathbf{m}$ and Σ_m are the prior mean and the prior covariance matrix respectively, and if the error $\mathbf{e}$ is distributed according to a Gaussian distribution with zero mean and known covariance matrix $e \sim \mathcal{N}(0, \Sigma_\mathbf{e})$, then the posterior distribution $\mathbf{m}|d \sim \mathcal{N}(\mu_{\mathbf{m}|\mathbf{d}}, \Sigma_{\mathbf{m}|\mathbf{d}})$ is a Gaussian distribution

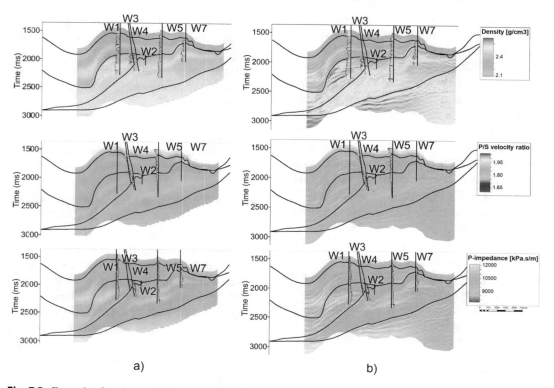

a)   b)

Fig. 7.3 Example of an deterministic elastic inversion. **a** Vertical well-section from the low-frequency models used in the inversion, **b** the inverted density, V_P/V_S and I_P models

with an analytical expression for the posterior mean $\mu_{\mathbf{m|d}}$ and the posterior covariance matrix $\Sigma_{\mathbf{m|d}}$ (Tarantola 2005):

$$\mu_{\mathbf{m|d}} = \mu_{\mathbf{m}} + \Sigma_{\mathbf{m}} \mathbf{F}^{\mathrm{T}} \left(\mathbf{F}\Sigma_{\mathbf{m}}\mathbf{F}^{\mathrm{T}} + \Sigma_{\mathbf{e}}\right)^{-1}(\mathbf{d} - \mathbf{F}\mu_{\mathbf{m}}),$$
(7.4)

$$\Sigma_{\mathbf{m|d}} = \Sigma_{\mathbf{m}} - \Sigma_{\mathbf{m}} \mathbf{F}^{\mathrm{T}} \left(\mathbf{F}\Sigma_{\mathbf{m}}\mathbf{F}^{\mathrm{T}} + \Sigma_{\mathbf{e}}\right)^{-1}\mathbf{F}\Sigma_{\mathbf{m}},$$
(7.5)

where $\mathbf{F}$ is the matrix that describes the linear forward operator.

These types of inversion methods have increased in importance in the industry since the pioneering work of Buland and Omre (2003), which was later extended for the multi-Gaussian case by Grana and Della Rossa (2010), for linearized rock physics inversion (e.g., Grana 2016; Grana et al. 2017) and linearized rock physics model in a multivariate multimodal prior distribution (e.g., de Figueiredo et al. 2018).

The great advantage of these types of seismic inversion methods is their ability to mathematically track the maximum a posteriori model (MAP) parameters and the corresponding standard deviation (Buland and Omre 2003). On the other hand, due to the assumption about model distribution and the linearization of the forward model they lack real exploration of the uncertainty space in lieu of exact prior information. To generate model realizations, the posterior distribution can then be sampled to generate model realizations, which represent alternative model solutions that fit equally well the observed seismic data (e.g., de Figueiredo et al. 2019).

When the Gaussian and linearized assumptions are not valid the posterior distribution can be approximated numerically with Markov chain Monte Carlo (McMC) types of methods. Sampling the posterior distribution might be non-trivial, requires expertise in the parameterization of the Markov chains and is in general computationally expensive. Recent developments on machine learning have been proposed to increase the performance of these sampling methods and alleviate the computational costs (e.g., de

Figueiredo et al. 2018; Aleardi 2020, 2021; Aleardi and Salusti 2021).

7.5   Geostatistical Seismic Inversion

Alternatively, stochastic sampling and optimization methods might be used to solve the seismic inversion problem without requiring the linearization of the forward model or assumptions about the distribution of the model parameters. In stochastic seismic inversion methods, the subsurface models ($\mathbf{m}$) are perturbed using stochastic optimization techniques such as simulated annealing, genetic algorithms, neighbourhood algorithm and geostatistical simulation and co-simulation (e.g., Sen and Stoffa 1991; Bortoli et al. 1993; Sambridge 1999; Le Ravalec-Dupin and Noetinger 2002; Soares et al. 2007; González et al. 2008; Grana et al. 2012; Azevedo et al. 2015, 2018; Azevedo and Soares 2017; Grana et al. 2021). When the perturbation of the model parameters requires the spatial coupling of the subsurface property (i.e., require the integration of a spatial covariance matrix) geostatistical simulation (Deutsch and Journel 1992) provides the set of tools that might be used as model perturbation technique and update. Below we describe two geostatistical seismic inversion methods.

7.5.1   Trace-by-Trace Geostatistical Inversion

Bortoli et al. (1993), and later Haas and Dubrule (1994) and Grijalba-Cuenca et al. (2000), introduced the pioneering works on geostatistical seismic inversion proposing a sequential trace-by-trace approach to invert full-stack seismic for I_P. Each seismic trace within the inversion grid (i.e., CMP location) is visited individually following the framework of a stochastic sequential simulation (Deutsch and Journel 1992). At each step along the pre-defined random path, Ns realizations of acoustic impedance are simulated using sequential Gaussian simulation (Deutsch and Journel 1992) taking the well-log data and previously visited/simulated traces into account as conditioning data and a variogram model describing the expected spatial continuity pattern.

A variogram model, or spatial covariance matrix, is a mathematical tool that describes the expected spatial continuity pattern of the property in study (e.g., Deutsch and Journel 1992; Goovaerts 1997). Modelling an experimental variogram is an extremely important step for any geostatistical inversion. An experimental variogram (γ) is first computed for different distances, $\mathbf{h}$, within the inversion grid (Eq. 7.6) using the available well-log data of the property of interest ($Z(x)$) located in x_α:

$$\gamma(\mathbf{h}) = \frac{1}{2N} \sum_{\alpha=1}^{N} [Z(x_\alpha) - Z(x_\alpha + \mathbf{h})]^2. \quad (7.6)$$

Then, we adjust a smooth, positive definite function defined by a reduced number of parameters (i.e., type of variogram model, range and sill) that describe the spatial continuity of $Z(x)$.

Modelling an experimental variogram is of utmost importance within the geostatistical framework, since it allows the synthesis of the structural characteristics of the spatial phenomena (e.g., degree of dispersion/continuity and anisotropies) into a single and coherent mathematical model. It is also common to adjust a model to the experimental variogram by conditioning it from expert knowledge about the phenomena being modelled (e.g., Goovaerts 1997).

In practice, for seismic reservoir characterization studies, the vertical variogram model is fitted to an experimental variogram computed from the well-log data. However, as the number of available wells is often reduced the calculation of horizonal experimental variograms is difficult. This is often the case in early exploration scenarios. In these cases, it is common to use the seismic amplitudes extracted at a reference horizon within the target area to compute the experimental variogram. As the seismic

amplitudes are smooth representations of the true subsurface this procedure results in an overestimation of the spatial continuities, which should be reduced for the inversion. Several alternative methods can be used to optimize these variogram parameters. A solution might be using a mini-inversion volume to test alternative scenarios and assess the impact of the different spatial continuity models in the predicted subsurface models. The preferable set of parameters can then be used to invert the entire seismic volume.

On trace-by-trace geostatistical seismic inversion, from the simulated impedance traces the corresponding reflection coefficients are computed and convolved with a wavelet. This step results in a set of Ns synthetic seismic traces. Each synthetic trace is compared against the observed seismic trace for the same location given an objective function, such as Pearson's correlation coefficient. The acoustic impedance realization that produces the lowest mismatch between the real and the synthetic seismic traces is integrated in the inversion grid as conditioning data for the simulation of the next location following the pre-defined random path. Different runs produce alternative inverted models as the random path changes and, consequently, modify the conditioning data at each trace location. The variability (e.g., the pointwise variance computed from an ensemble of methods) within the set of inverted models can be used to assess the spatial uncertainty related to the predicted property.

It is relevant to highlight that in trace-by-trace geostatistical seismic inversion methods, the algorithm will force a convergence in areas of the observed seismic volume associated with low signal-to-noise ratio. The sequential trace-by-trace approaches impose inverted models that fit the observed noisy seismic reflection data and as the simulated trace is assumed true data for the next CMP location, this can lead to the spread of the noisy (or unreliable) values.

Recently, Connolly and Hughes (2016) introduced a trace-by-trace approach to invert seismic data directly for subsurface rock properties, including facies, based on the concept of pseudo-wells. Each pseudo-well is a one-dimensional stochastic simulation of a stratigraphic profile (i.e., facies) with information about the subsurface physical properties. However, there are no explicit lateral spatial constraints and each trace is simulated independently. At each location within the inversion grid, many pseudo-wells are simulated with continuous-time Markov chains given a priori information regarding facies proportions. Those that give the best match are retained and considered good estimates of the reservoir properties. These models represent the posterior distributions of the predictions. The spatial continuity of the inverted rock and elastic properties is ensured in a second step with geostatistical simulation conditioned to models inverted in the previous step. As each location is simulated independently one of the advantages of this method is its potential to be highly parallelized.

7.5.2   Global Geostatistical Inversion

Global geostatistical seismic inversion methods use geostatistical methods, namely, stochastic sequential simulation and co-simulations (Deutsch and Journel 1992) as model perturbation technique. A detailed description of the basis on global iterative geostatistical inversion methods can be found in Azevedo and Soares (2017). Depending on the observed data, the existing inversion methods might predict the spatial distribution of elastic properties such as I_P (Soares et al. 2007) I_S (Nunes et al. 2012) and ρ, V_P and V_s (Azevedo et al. 2018) or be inverted directly for rock properties like facies, φ, fluid saturation and mineral fraction (Azevedo et al. 2019, 2020). The former requires the integration of a rock physics model to link the rock with the elastic domains (Mavko et al. 2009) (Chap. 8).

Global iterative geostatistical seismic inversion methods are based on two main ideas common to the different available methods: (i) direct sequential simulation and co-simulation (Soares 2001; Horta and Soares 2010; Soares et al. 2017) are the preferable methods to generate sets of models at each iteration; and (ii) the mismatch between observed and real seismic

traces drives the convergence of the iterative procedure.

In geostatistical elastic inversion (Fig. 7.5), the first iteration starts with the simulation of a set of Ns models, or geostatistical realizations, of the elastic properties of interest given the existing well-log data and a given variogram model. Before being included as conditioning data into the inversion grid, the well-log data needs to be upscaled. Backus averaging (Chap. 6) (Lindsay and Koughnet 2001) is the preferable upscaling tool as it describes the average V_P and V_S of a series of thin sedimentary layers. As best practice, after upscaling, the histograms of the original and upscaled logs should be compared. The upscaled histogram should reproduce the original one as it will be exactly reproduced in all the realizations generated during the iterative geostatistical seismic inversion. As the well data are often biased towards the pay-formations, depending on the complexity of the geological setting, there might be the need to extend the conditioning data set (i.e., the experimental data of the geostatistical simulation) to geological conditions not sampled, or poorly sampled, by the well data by Monte Carlo simulation (e.g., Avseth et al. 2005).

These realizations are then forward modelled to generate Ns synthetic seismic volumes. Depending on the observed data the forward model might comprise the calculation of the normal incidence or angle-dependent reflection coefficients, which are then convolved with a wavelet (Eq. 7.7):

$$\mathbf{A} = \mathbf{r} * \mathbf{w}, \qquad (7.7)$$

where the seismic amplitude ($\mathbf{A}$) represents the convolution of the reflection coefficients ($\mathbf{r}$) with a known wavelet ($\mathbf{w}$). The reflection coefficients calculation depends on the subsurface elastic properties (i.e., rock ρ and V_P and V_s).

Synthetic and real seismic are compared on a trace-by-trace basis considering the waveform and amplitude content of the signal. A possible equation that might be used as objective function and that is sensitive to both parameters is shown below:

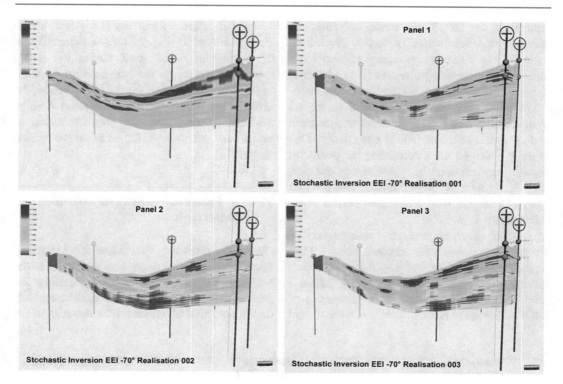

Fig. 7.4 (Top left) Seismic coloured impedance property from EEI workflow within a reservoir cellular model. This is compared to three geostatistical realization of impedance generated from a stochastic inversion (Top right, Bottom left, Bottom right). Note the higher level of detail and variability compared to the coloured result (Francis 2013). (Courtesy if GEO ExPro)

$$S = \frac{2\sum_{k=-n}^{n}\left(d_{t+k}d_{t+k}^{*}\right)}{\sum_{k=-n}^{n}(d_{t+k})^{2} + \sum_{k=-n}^{n}\left(d_{t+k}^{*}\right)^{2}}, \quad (7.8)$$

where d_t and d_t^* are the observed and predicted seismic traces at sample **t**, and **n** is the size of a moving window used to compute the similarity between both traces. As in Person's correlation coefficient, S varies between -1 and 1, for implementation purposes the negative values are truncated at zero.

The elastic traces that produce the highest S at a given iteration are stored along with the similarity coefficients in a set of auxiliary volumes. These volumes will be used as secondary variables in the stochastic sequential co-simulation of a new set of models in the following iterations. Regions of the inversion grid where S is high will exhibit a similar spatial pattern as the auxiliary elastic volumes. On the contrary, regions

where S is small the auxiliary volumes will have low impact. In other words, S controls the degree of assimilation of the observed data in the generation of elastic models in the subsequent iteration. The iterative procedure is over when a given number of pre-defined iterations is reached or when the global similarity between observed and predicted data is above a user-defined threshold.

Once the convergence is reached, all the co-simulated models in the last iteration produce synthetic seismic highly similar to the observed one (Fig. 7.5). A usual way to illustrate the results of geostatistical inversion is by computing the pointwise average model of all the realizations that produce synthetic seismic with a S higher than a given threshold (Fig. 7.5e). The average model represents the most likely model and has a smoother spatial distribution than the single realizations. The average model is equivalent to the MAP model obtained in Bayesian linearized seismic inversion (Sect. 7.4).

In this type of seismic inversion methodology, areas of poor signal-to-noise ratio in the observed data or areas where the wavelet is not representative of the recorded seismic tend to remain unmatched during the entire iterative procedure and represent areas of higher uncertainty. This effect is observed when computing the pointwise standard deviation, or the inter-quantile distance, (Fig. 7.5f) from all the realizations that produce synthetic seismic with a S higher than a given threshold.

All the simulated and co-simulated models share some common characteristics. They reproduce exactly the well-log data at the well locations (Fig. 7.5), the marginal (Fig. 7.6) and joint (Fig. 7.7) distribution of the properties of interest (for geostatistical elastic inversion) and

the spatial continuity patterns as revealed by the variogram models (Fig. 7.8). Also, depending on the variogram model used during the model simulation and due to the integration of the well-log data as conditioning data of the inversion, the inverted models show higher resolution than the recorded seismic, allowing characterization of layers at and below the seismic resolution (Fig. 7.9).

7.6 A Practical Guide to Seismic Inversion

One of the most important ideas that we would like to convey is that there is no such thing as the best seismic inversion method. Selecting the seismic inversion method to be applied depends on the objective of the study, the maturity of the

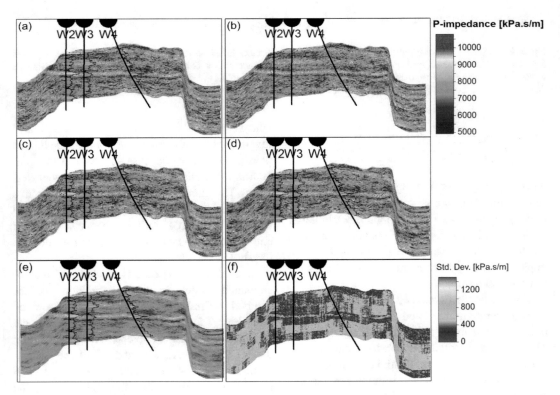

Fig. 7.5 Vertical well sections extracted from **a–d** four realizations of I_P generated during the last iteration of a geostatistical acoustic inversion; **e** pointwise average I_P model computed from all the realizations generated during the last iteration of a geostatistical acoustic inversion; **f** pointwise standard deviation of I_P computed from all the realizations generated during the last iteration of a geostatistical acoustic inversion

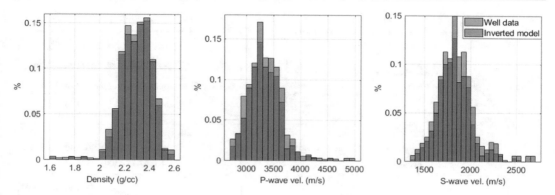

Fig. 7.6 Comparison between histograms computed from the well data used in a geostatistical seismic AVO inversion and the histograms computed from a set of realizations of density, P-wave and S-wave velocities

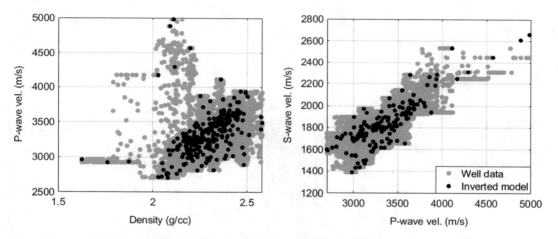

Fig. 7.7 Comparison between the joint distributions computed from the well data used in a geostatistical seismic AVO inversion and computed from a set of realizations. Joint distributions of density versus P-wave velocity and P-wave and S-wave velocity

sedimentary basin under exploration, the type of data available and their quality and the time and budget available to perform the study (Filippova et al. 2011; Francis 2006).

Nevertheless, and independent from the inversion method to be applied, there are a list of questions that should be taken into consideration when performing seismic reservoir characterization studies.

How large is the time window to be inverted? If there is already a target and this interval is relatively small (i.e., around hundreds of metres) geostatistical seismic inversion methods might be useful. If this area is large, alternative seismic

inversion methods might be preferable as with the increase of the time window the stationary assumption underlying geostatistical methods might not be valid. Within a geostatistical setting, an option is to perform the regionalization of the inversion grid along for example seismic units and assume the stationarity of the statistical properties per region (e.g., Nunes et al. 2017).

For deterministic inversion methodologies a large inversion window can also be problematic because we might face amplitude attenuation, loss of high frequencies and phase changes in the input seismic data. As discussed in Chap. 5, vertically and laterally varying wavelets along

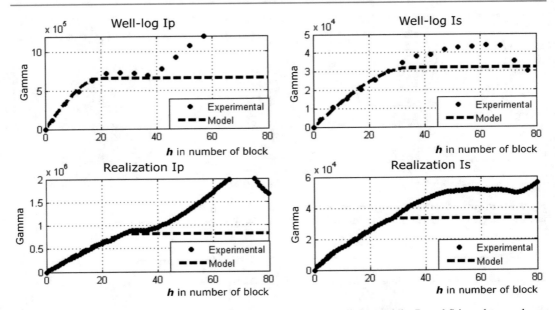

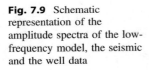

Fig. 7.8 Comparison between variogram models computed from the well data and the P- and S-impedance volumes obtained from a geostatistical elastic inversion

Fig. 7.9 Schematic representation of the amplitude spectra of the low-frequency model, the seismic and the well data

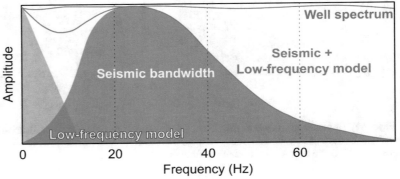

with Q-wavelets can overcome these issues but bring more complexity and possible uncertainty to the final result. Adequate vertical well control may also become a problem with large time windows. Starting models and low-frequency models created without full well data coverage from top to bottom of the inversion model may need to either extrapolate well values or use seismic velocities to fill data gaps. A more practical solution for deterministic approaches when dealing with large time window is to simply divide the window into two separate inversions.

In addition to the time window, decisions have to be made regarding the type of layering of the inversion model. Do we know enough information about the depositional environment? If so, we can use a stratigraphic grid that mimics this geological setting. On the other hand, we might opt for a simpler grid with parallel layering to the top and base of the inversion model. For deterministic inversions both the starting model and low-frequency model contain geological boundaries. The accuracy of these boundaries is important and will impact the predicted rock properties values. It is always important to understand the influence of model boundaries on the final inversion results and to avoid them overprinting the data-driven result we wish to obtain.

If during a feasibility study cross-plotting well-log rock physics data has been used to show the potential of seismic inversion for separating lithologies or fluids, it is important to see if the inverted volumes are able to reproduce this relationship. This can be easily done by extracting inversion values along pseudo-wells, and reproducing the cross-plots to assess if rock property discriminations match those from the feasibility study. Visual comparisons of pseudo-logs and well-logs can also be informative to validate the accuracy of the inversion.

A key parameter that links the geophysical and the geological domains is the wavelet (Chap. 5). The geoscientist should question the origin of the wavelet and its parameters such as the length, phase and polarity. Were there any wells involved during the wavelet extraction and how is the seismic-to-well tie? Does the synthetic at the well location match the record seismic trace? How is the reproduction of the reflectivity at the reservoir area? Again, how large is the inversion grid? Using a single wavelet assumes that the signal is spatially stationary. However, the wavelet does change its shape as the signal travels within the subsurface. For larger areas or seismic volumes with highly variable signal-to-noise ratio spatially dependent wavelets should be considered.

Most inversions whether determinist or geostatistical produce a range of volumes for QC purposes. These range in complexity and amount and can often overwhelm a non-specialist when faced with multiple volumes to review. Two of the most useful volumes to give the geoscientist an insight into the inversion quality are the synthetic volume(s) and residual volume(s). The synthetics are the inversions attempt to reproduce the input seismic based on the layer properties from the final results, with the residuals showing the mismatch between the inversion synthetics and input seismic. A map of residuals at key stratigraphic intervals can be very informative and may, in some cases, lead to refinements of

wavelet scaling, seismic conditioning and geological models.

Finally, a last word regarding the well data availability. After well-log data processing these data should be evaluated and a detailed exploratory data analysis performed. This evaluation should be done both at the original well-log resolution, seismic resolution and after upscaling into the inversion grid, which might be at the seismic resolution or at a higher resolution. The processed well-log data are also needed to model the expected spatial continuity pattern of the property to be inverted. This is often done by inferring a variogram model from the data. If needed, this matrix should include the a priori geological knowledge of the study area. In practical terms this implies modifying the spatial covariance matrix retrieved from the data according to these beliefs. Variogram models are critical for geostatistical seismic inversion methods as they control the variability (i.e., spatial and temporal resolution) of the inverted elastic and/or rock properties. Variogram modelling should always be carried out with realistic parameters relating to the true spatial spread of available well data and the size of the area of interest.

References

Aleardi M (2020) Combining discrete cosine transform and convolutional neural networks to speed up the Hamiltonian Monte Carlo inversion of pre-stack seismic data. Geophys Prospect 68(9):2738–2761

Aleardi M (2021) A gradient-based Markov chain Monte Carlo algorithm for elastic pre-stack inversion with data and model space reduction. Geophys Prospect 69 (5):926–948

Aleardi M, Salusti A (2021) Elastic pre-stack inversion through discrete cosine transform reparameterization and convolutional neural networks. Geophysics 86(1): R129–R146

Avseth P, Mukerji T, Mavko G (2005) Quantitative seismic interpretation: applying rock physics tools to reduce interpretation risk. Cambridge University Press, Cambridge

Azevedo L, Soares A (2017) Geostatistical methods for reservoir geophysics. Springer

Azevedo L, Nunes R, Soares A, Mundin EC, Neto GS (2015) Integration of well data into geostatistical seismic amplitude variation with angle inversion for facies estimation. Geophysics 80(6):M113–M128

Azevedo L, Nunes R, Soares A, Neto GS, Martins TS (2018) Geostatistical seismic amplitude-versus-angle inversion. Geophys Prospect 66(S1):116–131

Azevedo L, Grana D, Amaro C (2019) Geostatistical rock physics AVA inversion. Geophys J Int 216(3):1728–1739

Azevedo L, Grana D, Figueiredo L (2020) Stochastic perturbation optimization for discrete-continuous inverse problems. Geophysics 85(5):M73–M83. https://doi.org/10.1190/geo2019-0520.1

Bortoli LJ, Alabert F, Journel A (1993) Constraining stochastic images to seismic data. In: Soares A (ed) Geostatistics Troia'92. Kluwer, Dordrecht, pp 325–337

Bosch M, Mukerji T, Gonzalez EF (2010) Seismic inversion for reservoir properties combining statistical rock physics and geostatistics: a review. Geophysics 75(5):75A165-75A176

Buland A, Omre H (2003) Bayesian linearized AVO inversion. Geophysics 68(1):185–198

Castagna JP, Batzle ML, Eastwood RL (1985) Relationships between compressional-wave and shear-wave velocities in clastic silicate rocks. Geophysics 50:571–581

Connolly PA, Hughes MJ (2016) Stochastic inversion by matching to large numbers of pseudo-wells. Geophysics 81(2):M7–M22

de Figueiredo LP, Grana D, Bordignon FL, Santos M, Roisenberg M, Rodrigues BB (2018) Joint Bayesian inversion based on rock-physics prior modeling for the estimation of spatially correlated reservoir properties. Geophysics 83(5):M49–M61

de Figueiredo LP, Grana D, Roisenberg M, Rodrigues BB (2019) Multimodal Markov chain Monte Carlo method for nonlinear petrophysical seismic inversion. Geophysics 84(5):M1–M13

Deutsch C, Journel AG (1992) GSLIB: geostatistical software library and users' guide. Oxford University Press, New York

Doyen P (2007) Seismic reservoir characterization: an Earth modelling perspective. EAGE

Filippova K, Kozhenkov A, Alabushin A (2011) Seismic inversion techniques: choice and benefits. First Break 29(5):103–114

Francis AM (2006) Understanding stochastic inversion: part 1. First Break 24(11):79–84

Francis A (2013) Geophysics: A Simple Guide to Seismic Inversion. GEO ExPro 10(2)

Gardner GHF, Gardner LW, Gregory AR (1974) Formation velocity and density—the diagnostic basics for stratigraphic traps. Geophysics 39:770–780

González EF, Mukerji T, Mavko G (2008) Seismic inversion combining rock physics and multiple-point geostatistics. Geophysics 73(1):R11–R21. https://doi.org/10.1190/1.2803748

Goovaerts P (1997) Geostatistics for natural resources evaluation. Oxford University Press, New York

Grana D, Della Rossa E (2010) Probabilistic petrophysical-properties estimation integrating statistical rock physics with seismic inversion. Geophysics 75(3):O21–O37

Grana D, Mukerji T, Dvorkin J, Mavko G (2012) Stochastic inversion of facies from seismic data based on sequential simulations and probability perturbation method. Geophysics 77(4):M53–M72

Grana D (2016) Bayesian linearized rock-physics inversion. Geophysics 81(6):D625–D641. https://doi.org/10.1190/GEO2016-0161.1

Grana D, Fjeldstad T, Omre H (2017) Bayesian Gaussian mixture linear inversion for geophysical inverse problems. Math Geosci 49:1–37

Grana D, Mukerji T, Doyen P (2021). Seismic reservoir modeling: theory, examples and algorithms. Wiley-Blackwell

Grijalba-Cuenca A, Torres-Verdin C, van der Made P (2000) Geostatistical inversion of 3D seismic data to extrapolate wireline petrophysical variables laterally away from the well. SPE Annual technical conference and exhibition, SPE-63283-MS, Dallas, Texas

Haas A, Dubrule O (1994) Geostatistical inversion—a sequential method of stochastic reservoir modelling constrained by seismic data. First Break 12(11):561–569

Hampson D, Russell B (1985) Maximum-likelihood seismic inversion. In: Proceeding of the 12th annual national Canadian Society of exploration geophysicists (CSEG) meeting, abstract no. SP-16, Calgary, Alberta

Horta A, Soares A (2010) Direct sequential co-simulation with joint probability distributions. Math Geosci 42(3):269–292. https://doi.org/10.1007/s11004-010-9265-x

Lancaster S, Whitcombe D (2000) Fast-track 'coloured' inversion. SEG expanded abstracts, 3–6.

Le Ravalec-Dupin M, Noetinger B (2002) Optimization with the gradual deformation method. Math Geol 34(2):125–142

Lindsay R, Koughnet RV (2001) Sequential Backus averaging: upscaling well logs to seismic wavelengths. Lead Edge 20(2):188–191. https://doi.org/10.1190/1.1438908

Martinho M, Tylor-Jones T, Azevedo L (2021) The impact of seismic inversion methodology on rock property prediction. Geophys Prospect 69(2):388–403

Mavko G, Mukerji T, Dvorkin J (2009) The rock physics handbook—tools for seismic analysis of porous media, 2nd edn. Cambridge University Press

Nunes R, Soares A, Azevedo L, Pereira P (2017) Geostatistical seismic inversion with direct sequential simulation and co-simulation with multi-local distribution functions. Math Geosci 49(5):583–601

Nunes R, Soares A, Neto GS, Dillon L, Guerreiro L, Caetano H, Maciel C, Leon F (2012) Geostatistical inversion of prestack seismic data. In: Ninth international geostatistics congress. Oslo, Norway, pp 1–8

Oldenburg DW, Scheuer T, Levy S (1983) Recovery of the acoustic impedance from reflection seismograms. Geophysics 48(10):1318–1337

Russell, B, Hampson, D (1991) Comparison of poststack seismic inversion methods. SEG Technical Program Expanded Abstracts, pp 876–878

Russell B (1988) Introduction to seismic inversion methods. SEG

Sambridge M (1999) Geophysical inversion with a neighbourhood algorithm—I. Searching a parameter space. Geophys J Int 138(2):479–494

Scales JA, Smith ML, Treitel S (2001) Introductory geophysical inverse theory. Samizdat Press

Sen MK, Stoffa PL (1991) Nonlinear one-dimensional seismic waveform inversion using simulated annealing. Geophysics 56(10):1624–1638

Soares A (2001) Direct sequential simulation and cosimulation. Math Geol 33(8):911–926

Soares A, Diet JD, Guerreiro L (2007) Stochastic Inversion with a global perturbation method. In: EAGE Conference on Petroleum Geostatistics, 10–14

Soares A, Nunes R, Azevedo L (2017) Integration of Uncertain Data in Geostatistical Modelling. Mathematical Geosciences 49: 253–273

Tarantola, A (2005) Inverse problem theory and methods for model parameter estimation. Society for Industrial and Applied Mathematics

Yilmaz O (1987) Seismic data processing. Society of Exploration Geophysicists, Tulsa

Rock Properties Prediction

The goal of a geoscientist is to create models of the subsurface rock properties to be used in fluid flow simulation and reserves calculation. The inverted elastic models resulting from seismic inversion (Chap. 7) are often used as a starting point for this step of the geo-modelling workflow. This approach allows deterministic inversion elastic rock property volumes to be used in a workflow which produces probability volumes for single or multiple lithology or fluid types. Having a high level of accuracy in the deterministic elastic properties is key to this workflow being successful. Careful QC of not only the elastic properties (e.g., acoustic impedance and Vp/Vs ratio), but also the probability volumes to well control must be carried out to avoid unplausible classifications. For this reason, this workflow is not recommended when well data are sparse or the geology is poorly understood. In fact, predicting rock properties from elastic models in a sequential two-step approach might lead to unplausible geological models (i.e., models that are not similar to the true subsurface geology) as these properties are not directly linked to the observed geophysical data. Also, and for the same reason, the sequential approach might not ensure a proper propagation of uncertainty throughout the entire geo-modelling workflow limiting our ability to assess risks.

Alternatively, rock properties, including facies, might be directly predicted from seismic reflection data. During the writing of this book this topic is still an important research topic in the geophysical community. These seismic inversion methods are often designated as single-loop inversion methods and at some point require the integration of rock physics models (Mavko et al. 2009) within the inversion workflow. In principle, seismic rock physics inversion methods allow for more reliable uncertainty assessment of the predicted properties (Bosch 2004; Bosch et al. 2010). Figure 8.1 exemplifies two different inversion workflows of sequential and single-loop facies prediction from seismic data assuming a geostatistical seismic inversion framework (Sect. 7.5). Alternative geo-modelling workflows could be used based on the product of deterministic seismic inversion methods.

While in the next sections we focus on geostatistical seismic rock physics inversion methods where the model perturbation and update are performed with stochastic sequential simulation and co-simulation (Deutsch and Journel 1992), alternative methods for simultaneous elastic and rock property prediction exist. The following list of references comprise the most relevant works published at this time regarding this topic (e.g., Kemper and Gunning 2014; Connolly and Hughes 2016; Fjeldstad and Omre 2020; de Figueiredo et al. 2019).

When rock properties are predicted from inverted elastic models a first step is often the classification of the inverted models into facies (or lithofluid facies) (Fig. 8.2). In single-loop approaches such as rock physics seismic inversion method, it is also common to have a facies

T. Tylor-Jones and L. Azevedo, *A Practical Guide to Seismic Reservoir Characterization*, Advances in Oil and Gas Exploration & Production, https://doi.org/10.1007/978-3-030-99854-7_8

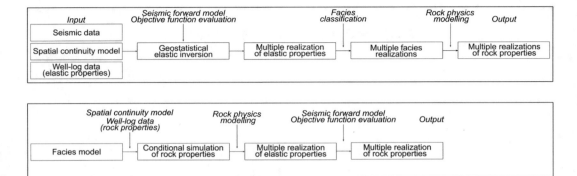

Fig. 8.1 Comparison between sequential and single-loop geostatistical seismic inversion methods

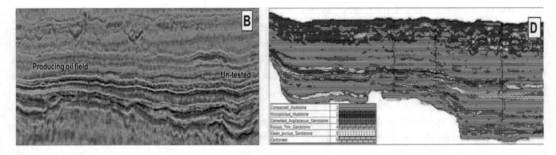

Fig. 8.2 Examples of fluid classification (Left) and facies classification (Right) coming from inverted acoustic impedance and Vp/Vs ratio using well based facies classification (KjØlhamar et al. 2020) (Courtesy of TGS)

classification. For these reasons, in the following section, we briefly introduce Bayesian Naïve Classification and Expectation–Maximization (e.g., Avseth et al. 2005; Hastie et al. 2009). Grana et al. (2017) shows a detailed comparison of the results obtained by both methods.

8.1   Facies Classification Methods

The spatial distribution of facies can be predicted by classifying seismic attributes (i.e., elastic and/or rock properties) using different methods. Statistical-based classification methods are preferable when comparing to those based purely on deterministic approaches due to their ability to account for uncertainty. In this section, we do not provide an extensive overview on classification algorithms, but we focus on two algorithms that have been widely used in subsurface modelling

and characterization: Bayesian Naïve Classification (e.g., Avseth et al. 2005; Hastie et al. 2009) and Expectation–Maximization (EM) (Hastie et al. 2009).

Bayesian classification uses a training data set built from the well-log data and the collocated well-log facies profile. At the well location, the well-log curves of interest, such as porosity, mineral volume fraction and fluid saturation, are classified into a pre-defined number of N_f facies. N_f depends on the objective of the study and the geological background. For reservoir characterization, examples of facies include shale, brine sand and hydrocarbon sand (Fig. 8.3). The statistical properties (i.e., mean, covariance and proportions), as inferred from the training data, are used to compute the prior and likelihood function for the Bayesian classification according to Bayes' rule:

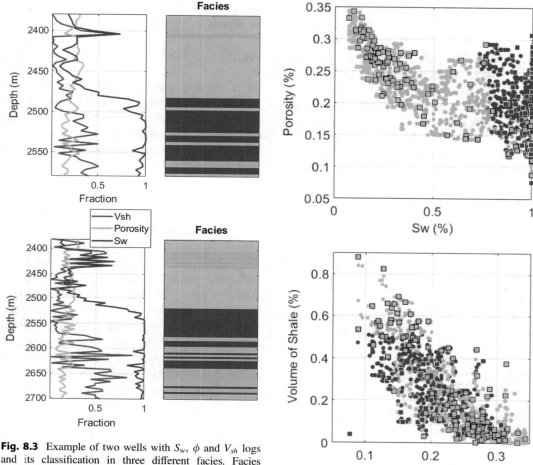

Fig. 8.3 Example of two wells with S_w, ϕ and V_{sh} logs and its classification in three different facies. Facies classified as shales are plotted in grey, as brine sands in blue and as oil sands in green

Fig. 8.4 Relationship between S_w versus ϕ **a** and **b** ϕ versus V_{sh} coloured by facies. Bayesian Naïve Classification was used as classifier. Classified well-log samples are plotted as squares and geostatistical realization in circles. Facies classified as shales are plotted in grey, as brine sands in blue and as oil sands in green

$$P(k|\boldsymbol{d}) = \frac{P(\boldsymbol{d}|k)P(k)}{P(\boldsymbol{d})} = \frac{P(\boldsymbol{d}|k)P(k)}{\sum_{k=1}^{N_f} P(\boldsymbol{d}|k)P(k)}, k$$
$$= 1, \ldots, N_f,$$

(8.1)

where $\boldsymbol{d}$ is the vector of the petrophysical properties used for the classification, the previously simulated petrophysical models, and k is the facies value. In Eq. 8.1, $P(\boldsymbol{d}|k)$ is the likelihood function, $P(k)$ is the prior model and $P(\boldsymbol{d})$ is a normalization constant.

An example of the facies classification from the well-log data and its application in set of rock models calculated with geostatistical simulation is shown in Fig. 8.4.

While Bayesian Naïve Classification is a supervised method, the EM classification algorithm is unsupervised. This means that the user does not need to explicitly model the facies from the well-log data but just provide the number of expected classes (i.e., N_f). The EM algorithm is an iterative method for finding the maximum-likelihood estimates of parameters in statistical models, where the model depends on unobserved latent variables (i.e., the facies in the context of subsurface modelling and characterization).

We start by modelling the distribution of the input data points through a multimodal model d, which can be described as a mixture of N_f Gaussian components (i.e., a linear combination of Gaussian distributions) (Grana and Della Rossa 2010). The set of data points in our examples are, for example, inverted models from seismic data (e.g., V_P, V_S, S_w) or well-log data. The goal in EM classification is to fit the model to the data by maximum likelihood. The parameters that need to be estimated are as follows:

$$\boldsymbol{\theta} = \left(\pi_1, \ldots, \pi_{N_f-1}, \boldsymbol{\mu}_{d|1}, \ldots, \boldsymbol{\mu}_{d|N_f-1}, \boldsymbol{\Sigma}_{d|1}, \ldots, \boldsymbol{\Sigma}_{d|N_f} \right),$$
(8.2)

where π is the proportion of each facies (i.e., the weight of each Gaussian component), $\boldsymbol{\mu}$ and $\sum$ are the mean and covariance matrix of d in each facies.

We can define the log-likelihood $(l(\boldsymbol{\theta}, \boldsymbol{d}))$ based on the n_s measured data points as

$$\begin{aligned} l(\boldsymbol{\theta}, \boldsymbol{d}) &= \sum_{j=1}^{n_s} \sum_{k=1}^{N_F} \log\left(\pi_k f_{d|k}(\boldsymbol{d}_j) \right) \\ &= \sum_{j=1}^{n_s} \sum_{k=1}^{N_F} \log\left(\pi_k N_{n_d}\left(\boldsymbol{d}_j; \boldsymbol{\mu}_{d|k}, \boldsymbol{\Sigma}_{d|k} \right) \right). \end{aligned}$$
(8.3)

The objective is then to estimate the set of parameters $\boldsymbol{\theta}$ that maximizes the log-likelihood function $l(\boldsymbol{\theta}, \boldsymbol{d})$:

$$\boldsymbol{\theta}_{ML} = argmax_{\theta} l(\boldsymbol{\theta}, \boldsymbol{d}).$$
(8.4)

Maximizing $l(\boldsymbol{\theta}, \boldsymbol{d})$ analytically is difficult and costly. Alternatively, numerical approaches can be used (Hastie et al. 2009). This estimation can be performed sequentially with two steps: the expectation and the maximization steps.

For the expectation step, we first consider an indicator variable $\boldsymbol{k} = \left[k_1, \ldots, k_{N_f} \right]^T$, where for a given $j \in \{1, \ldots, n_s\}$, k_i takes the value 1 if $\boldsymbol{d}_j$ comes from $f_{d|i}(\boldsymbol{d}_j)$ and 0 otherwise. As in the classification of three-dimensional subsurface models, we do not know the values of the

indicator beforehand, we substitute each $k_{k,j}$ by its expected value E:

$$\gamma_{k,j}(\boldsymbol{\theta}) = E\left(k_{k,j} | \boldsymbol{\theta}, \boldsymbol{d}_j \right) = p\left(k_k = 1 | \boldsymbol{\theta}, \boldsymbol{d}_j \right). \quad (8.5)$$

Then, we compute the maximization step. The expectations computed in the previous step are used to calculate the weighted means and covariances to update the estimates of the parameters. These steps are then repeated until convergence.

8.2   Single-Loop Inversion for Facies and Rock Properties

We present two alternative methods for facies and rock properties prediction from seismic reflection data. The first is an iterative geostatistical rock physics AVO inversion (Azevedo et al. 2019; Cyz and Azevedo 2020), which can be considered an extension of the iterative geostatistical seismic inversion methods shown in Sect. 7.5 and includes a facies classification step within the inversion workflow. The second is a stochastic optimization inversion method for the simultaneous prediction of facies and rock properties (Azevedo et al. 2020). In the latter, facies are perturbed and updated iteratively and directly from the data mismatch.

The iterative geostatistical rock physics AVO inversion aims at predicting subsurface rock properties from pre- or partially seismic data. It is divided into four main steps: (i) petrophysical model generation by stochastic sequential simulation and co-simulation. Depending on the objective of the inversion these properties can be porosity (ϕ), volume of shale (V_{sh}) and water saturation (S_w) (Azevedo et al. 2019) or brittleness index, total organic carbon, ϕ and volume of clay (Cyz and Azevedo 2020); (ii) facies classification based on the models generated in (i); (iii) elastic property calculation by applying a pre-calibrated facies-dependent rock physics model and synthetic seismic calculation; (iv) comparison between real and synthetic seismic data and generation of new petrophysical models for the next iteration constrained by data mismatch computed in (iii).

The first step consists of sequentially generating an ensemble of *Ns* models of rock properties using stochastic sequential simulation and co-simulation. Direct sequential simulation (Soares 2001) and direct sequential co-simulation with joint probability distributions (Horta and Soares 2010) are used as the model perturbation technique. Differently from Sequential Gaussian Simulation (SGS) (Deutsch and Journel 1992), the use of DSS allows the direct use of the distribution of the property to be simulated as estimated form the experimental data (i.e., the well-log data). This characteristic of direct sequential simulation is of interest when the distribution retrieved from the experimental well data is not Gaussian. The use of the experimental data in its original domain becomes more critical for co-simulation, as it allows reproducing of the joint behaviour of the primary and secondary variables of interest even when this relationship is complex. Reproducing the marginal distribution of each property and joint behaviour of pairs of properties is important to ensure geological consistency of the resulting models and the reliability of the facies classification step.

The simulated rock properties are then classified into *Ns* facies models. We use a Bayesian classification approach (Sect. 1.1) based on the well-log data and the collocated well-log facies profile. Alternative classification methods can be used.

Based on the available well-log data we first calibrate a rock physics model that fits the observed data within each facies. Several rock physics models can be used; however, the general rock physics workflow aims at predicting the elastic response including ρ, V_P and V_S when the petrophysical properties are known.

Once the elastic response is calculated, angle-dependent synthetic seismic data can be obtained by computing angle-dependent reflection coefficients $\left(R_{pp}(\theta)\right)$ using, for example, three-term Shuey's linear approximation (Shuey 1985). Alternative and more exact methods to compute the reflection coefficients can be used (Table 6.2).

The resulting reflection coefficients are convolved with angle-dependent wavelets to generate synthetic seismic data.

The stochastic perturbation of the petrophysical models is based on a data selection procedure performed at the end of each iteration and based on the principle of crossover genetic algorithm. The best portions of the petrophysical models generated at a given iteration (i.e., the regions of the model with the highest match between real and synthetic seismic data) are used as seed for the generation of a new family of models during the next iteration.

After generating the angle-dependent synthetic seismic, each synthetic and real trace with a total number of N samples is compared and the mismatch is evaluated according to S (Eq. 7.8).

At each iteration, from the ensemble of rock properties models generated during the first stage of the proposed algorithm, we select the traces that ensure the highest S for all the angles simultaneously. These traces are used as secondary variables in the co-simulation of a new set of models during the following iteration.

The geostatistical rock physics seismic AVO inversion can be summarized in the following sequence of steps (Fig. 8.5).

(i) Generation of *Ns* rock property models with DSS and co-DSS conditioned to the available log data and to a spatial continuity pattern as revealed by a variogram model.

(ii) Lithofluid facies classification of the rock property models generated in (i), (ii) and (iii) into a number of pre-defined facies using, for example, Bayesian classification (Sect. 1.1).

(iii) Application of facies-dependent rock physics models in each facies to compute the elastic response of the models generated in (i), (ii) and (iii).

(iv) Calculation of synthetic seismic data.

(v) Comparison on a trace-by-trace basis of synthetic and real seismic data (S).

(vi) Selection from the models generated in (i), (ii) and (iii) of the portions that ensure the highest correlation coefficient simultan-

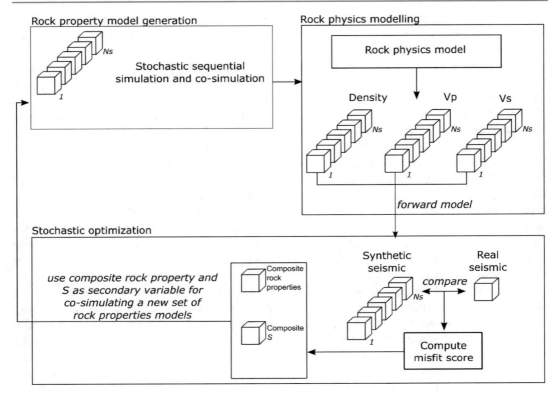

Fig. 8.5 Schematic representation of the iterative geostatistical seismic rock physics inversion

eously for all the angles at each reservoir location.

(vii) Generation of a new set of rock property models by using co-DSS with local correlation coefficients.

(viii) Iterate until a given global correlation coefficient between the angle-dependent synthetic and real seismic data is reached.

We illustrate this inversion methodology with an application example from a small portion of the Norne field (North Sea) (Bjørke 2009; Dadashpour 2009) (Fig. 8.6). The rock and elastic properties are predicted from a data set that includes three partial angle stacks with central angles of incidence of 10°, 22.5° and 35° (i.e., near-stack, mid-stack and far-stack, respectively) and four wells (Fig. 8.7). Only wells W1 and W3 have a complete log suite composed of P-wave and S-wave velocity,

density, water saturation, volume of shale and porosity logs.

The spatial continuity pattern of each petrophysical property was imposed through variogram models fitted to experimental variograms, computed from the upscaled-log data for the vertical direction and from the real full-stack volume for the horizontal direction. These were kept constant throughout the entire inversion and are reproduced in all realizations generated during the inversion.

Well-logs were used for well-log facies classification, rock physics calibration and as conditioning data for the generation of the reservoir models of water saturation, porosity and volume of shale. Three facies have been defined: shale with volume of shale higher than or equal to 40%; brine sand with volume of shale smaller than 40% and water saturation higher than or equal to 80% and oil sand with volume of shale smaller than 40% and water saturation smaller than 80%

Fig. 8.6 Seismic grid and relative location of wells for the application example of geostatistical rock physics inversion. The thin purple line represents the location of the vertical well section used to illustrate the results of the inversion

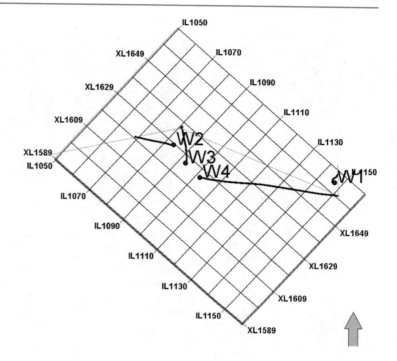

(Suman and Mukerji 2013). The elastic response in sandstone was modelled with the stiff-sand model. A mixture of quartz and feldspar ($K_{sa} = 25\,\text{GPa}$, $G_{sa} = 20\,\text{GPa}$, $\rho_{sa} = 2.64\,\text{g/cm}^3$) was considered for the sandstone phase. The matrix bulk and shear moduli were computed using Voigt-Reuss-Hill average. The critical porosity set as 0.49 and the effective fluid bulk modulus was computed using Batzle-Wang relation (Batzle and Wang 1992) and Reuss average. Gassmann's equation was applied to compute the fluid effect and predict the elastic behaviour in wet conditions. Han's model empirically calibrated using well-log data at the well location (e.g., Mavko et al. 2009) was used to model the elastic response in shale. The existing well-log samples of water saturation, porosity and volume of shale for wells W2 and W4 were used exclusively as conditioning data for the stochastic sequential simulation and co-simulation.

The inversion ran with 6 iterations with the generation of 32 realizations of water saturation, porosity and volume of shale per iteration. As the perturbation of the model parameter space is done with stochastic sequential simulation and co-simulation, the histograms of water saturation, porosity and volume of shale as inferred from the upscaled-log samples are reproduced in all models generated during the inversion. The reproduction of the joint distribution of water saturation and porosity and porosity and volume of shale is also ensured by using direct sequential co-simulation with joint probability distributions (Fig. 8.4).

The real partial angle stacks show a bright reflection around the 2400 ms corresponding to the top of the Not formation (Fig. 8.7a–c). The spatial continuity of the reflections decreases with depth, which makes the convergence of the inversion in these areas more challenging. Below 2600 ms it is hard to identify any spatially coherent reflection. As the incidence angle increases, the number of continuous reflections decreases. The far-stack volume (Fig. 8.7c) shows a single clear reflection corresponding to the top of the Not formation.

The predicted synthetic seismic reflection data computed from the best-fit inverted water saturation, porosity and volume of shale models are

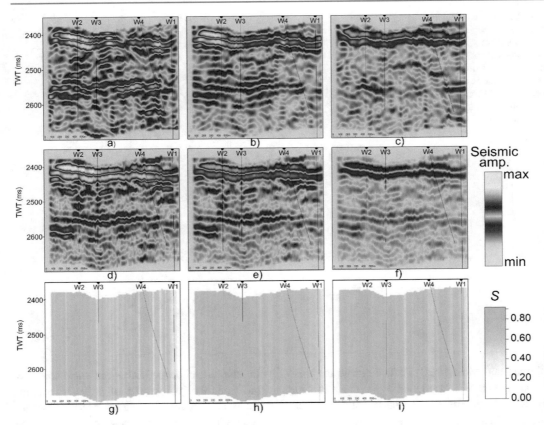

Fig. 8.7 Vertical well sections extracted from: **a** the real near-stack volume; **b** the real mid-stack volume; **c** the real far-stack volume; **d** the synthetic near-stack volume; **e** the synthetic mid-stack volume; **f** the synthetic far-stack volume; **g** the trace-by-trace S between real and synthetic near-stack volumes; **h** the trace-by-trace S coefficient computed between real and synthetic mid-stack volumes; **i** the trace-by-trace S between real and synthetic far-stack volumes

shown in Fig. 8.7e–f. The synthetic predicts the location and amplitude of the main seismic reflections. However, it introduces higher reflections continuity compared to the observed seismic. This is a common effect on synthetic seismic predicted by inversion methods. The synthetics often exhibit larger spatial continuity ranges. The difficulty in reproducing the bottom part of the seismic data (below 2600 ms) is shown by the lower trace-by-trace S coefficient computed between real and synthetic seismic traces (Fig. 8.7g–i). As a global geostatistical approach (i.e., the model perturbation is performed for the entire grid at once and not on a trace-by-trace basis), low signal-to-noise regions tend to be unmatched.

Figure 8.8a–c shows vertical well sections extracted from the pointwise average models of water saturation, porosity and volume of shale computed from all the realizations generated during the last iteration. These models show a larger spatial variability in the upper part of the inversion model associated with the regions of lower water saturation and preserve with the geological features interpreted from the original seismic.

We use wells W4 and W2 to verify locally the predicted elastic samples in the two wells not used for the calibration of the rock physics (Fig. 8.9). We compare the existing measured log samples of density, P-wave and S-wave velocity to the minimum, average and maximum

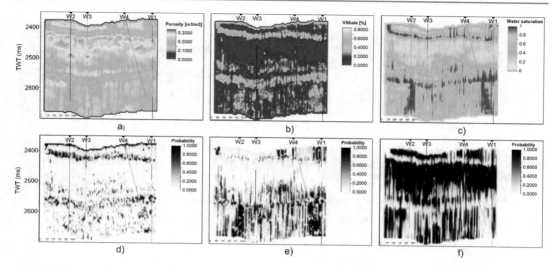

Fig. 8.8 Vertical well sections extracted from: **a** the pointwise average model of porosity; **b** the pointwise average model of volume of shale; **c** the pointwise average model of water saturation; **d** the probability of shale; **e** the probability of brine sand; **f** the probability of oil sand

values generated during the last iteration of the geostatistical inversion. For well W4 (top row Fig. 8.9), the density and V_P logs agree with the average predicted values and is mainly within the bounds of the minimum and maximum predicted values. Well W2 (bottom row Fig. 8.9) is correctly constrained for almost the entire extent of the well where all the models generate the same water saturation, porosity and volume of shale and consequently the same elastic response. The most likely facies model computed from all the facies volumes generated during the last iteration do match the log facies interpretation (fifth plot in each row of Fig. 8.9).

Finally, the facies probability volumes of shale, brine and oil sand are shown in Fig. 8.8. These are computed by using the ensemble of facies models classified from the water saturation, porosity and volume of shale models generated during the last iteration. It is possible to interpret a main oil sand region immediately below the top Not formation (~ 2400 ms) and a thinner one below the top A are formation (~ 2600 ms). As the facies models are classified from the simulated and co-simulated rock property models, these are not directly linked to the data mismatch and there is not direct constraint about the ordering of the facies there is the

possibility to get unphysical scenarios (e.g., brine sands on top of oil sands) in the predicted facies. Next, we illustrate one possibility to perturb and update facies directly from the seismic data using a similar geostatistical framework.

The joint prediction of facies and continuous properties, such as porosity, requires the joint optimization of both domains according to the data mismatch (Azevedo et al. 2020). As the facies constrain the spatial distribution of the rock and elastic properties, these should be the first to be generated. The inversion method illustrated here uses a one-dimensional Markov chain model (e.g., Krumbein and Dacey 1969; Elfeki and Dekking 2001; Lindberg and Grana 2015; Fjeldstad and Omre 2020) to impose a vertical correlation, preserve facies proportions, their vertical ordering and avoid implausible geological situations. Alternative approaches to generate facies models based on different modelling algorithms have been proposed by González et al. (2008) and Connolly and Hughes (2016).

In the Markov chain approach, facies are sequentially generated along the profile. The facies values belong to a finite number of N_f classes and the probability distribution of the facies (i.e., the chain state) at a given location is

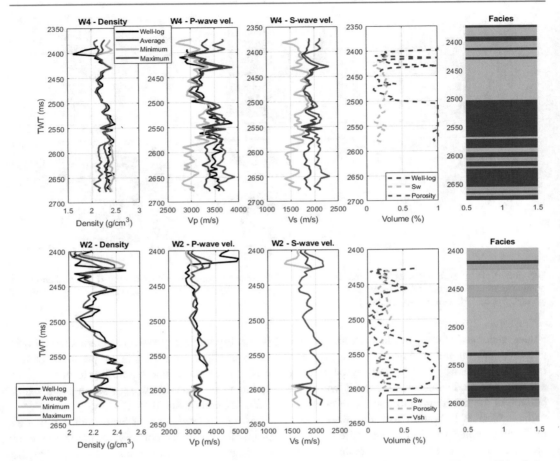

Fig. 8.9 Comparison between well-log and predicted minimum, average and maximum of density, V_P and V_S samples in the last iteration of the inversion. The upscaled-log samples used as conditioning data for both locations are shown in the fourth track, these do not cover the entire extent of the wells. The resulting facies classification is shown in the fifth track: samples classified as shales are plotted in grey, as brine sands in blue and as oil sands in green

conditioned by the facies at the location above, rather than the entire sequence of previously simulated facies. The probabilities of transitioning from one facies to another one (namely, the transition probabilities) are represented in a transition matrix, where each rows represent the facies at the previous location and the columns represent the facies at the current location. As discussed for the variogram models, the transition matrix might be inferred from the log facies profile at the well location.

The first sample, at the top of the profile, is simulated from the prior probability of the facies $P(f)$. Then, we sequentially simulate the following samples from the transition probabilities:

$$p_{i,j} = P(f_t = i | f_{t-1} = j), \qquad (8.6)$$

for $i, j = 1, \ldots, N_f$ and for $t = 2, \ldots, N_s$, where N_s is the number of samples in the profile. We then obtain a vector of facies samples $f^1 = \left[f_1^1, \ldots, f_{N_s}^1 \right]^T$, where the superscript 1 indicates the iteration number in the optimization. We independently repeat the same procedure for N_r realizations and obtain the ensemble of facies realizations $\{ f^{1,1}, \ldots, f^{1,N_r} \}$.

After simulating facies, the next step is the generation of the rock properties. To ensure that these properties are spatially correlated, stochastic sequential simulation and co-simulation methods can be used (Deutsch and Journel 1992). Within this framework, the rock properties are simulated conditioned to the previously simulated facies models, ensuring that marginal and joint probability distributions of rock properties and their correlations are reproduced in the simulated models. Each N_r facies model, generated with the first-order Markov chain in the previous step, is used as conditioning data for generating geostatistical realizations of the relevant rock properties under investigation (such as porosity ϕ, volume of shale V_{Sh} and water saturation S_w). Direct sequential simulation and co-simulation (Soares 2001) with multi-local distribution functions (Nunes et al. 2017) can be used at this step as they allow simulating petrophysical properties by preserving the marginal and joint distribution of the model variables as inferred from the well data to each facies.

For example, in petrophysical inversion, water saturation is first simulated conditioned by the facies, then porosity, and finally the volume of shale as in Azevedo et al. (2019):

$$S_w^{1,r} \sim F_{sw}(S_w | f^{1,r}), \qquad (8.7)$$

$$\phi^{1,r} \sim F_\phi(\phi | S_w^{1,r}, f^{1,r}), \qquad (8.8)$$

$$V_{Sh}^{1,r} \sim F_{vsh}(V_{Sh} | \phi^{1,r}, f^{1,r}), \qquad (8.9)$$

where F_{sw}, F_ϕ and F_{vsh} can be inferred from well-log data, for example.

Elastic properties can be predicted from pre-calibrated rock physics relations to compute density, P- and S-wave velocity models. For each set of N_r geostatistical realizations of petrophysical properties $m^{1,r}$, for $r = 1, \ldots, N_r$, we compute the corresponding N_r elastic models $l^{1,r} = k(m^{1,r})$, including impedances or velocities and densities, using a pre-calibrated deterministic rock physics model k, such as the stiff-sand model (Mavko et al. 2009). Synthetic seismic data are computed from the set of N_r elastic

model as $d^{1,r} = h(l^{1,r}, f^{1,r})$ for $r = 1, \ldots, N_r$, where h is the seismic convolutional model.

The rock properties that ensured the maximum S are stored in auxiliary volumes to be used in the subsequent iteration to generate new ensembles of models. The facies volume is used to update the facies probability distribution at each location using the probability perturbation method (Caers and Hoffman 2006; Grana et al. 2012). The remaining rock properties are co-ssimulated using the auxiliary volumes along with the local S coefficients to generate a new set of models. The iterative procedure proceeds until a fixed number of iterations is reached. The joint stochastic optimization preserves the correlation between the discrete and the continuous domains.

This inversion method is illustrated in the same application example with similar parameterization as before to highlight the difference when jointly optimizing facies and rock properties. One of the main differences is the need to estimate a vertical transition matrix. The well-log data was used to compute the following transition matrix:

$$T = \begin{bmatrix} 0.59 & 0.30 & 0.11 \\ 0.30 & 0.70 & 0 \\ 0.07 & 0.02 & 0.91 \end{bmatrix}. \qquad (8.10)$$

The average models of petrophysical properties and the facies probabilities computed from the ensemble of stochastic realizations generated during the last iteration are shown in Fig. 8.10a–c. In general, the predictions of each property match the measured ones at the well location while they exhibit lateral continuity even though the facies models are being simulated using exclusively a vertical correlation model. The lateral continuity is provided by the match between observed and predicted seismic data. However, these models still show some vertical striping originated by the trace-by-trace approach when simulating facies.

The facies probability volumes (Fig. 8.10d, e) show that the facies ordering as imposed by the transition matrix (Eq. 8.10) is preserved during

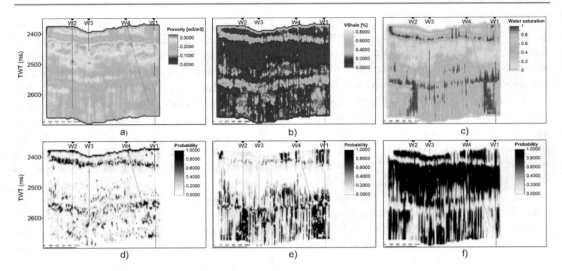

a)   b)   c)

d)   e)   f)

Fig. 8.10 Vertical well sections extracted from the simultaneous optimization of continuous and facies models results: **a** the pointwise average model of porosity; **b** the pointwise average model of volume of shale; **c** the pointwise average model of water saturation; **d** the probability of shale; **e** the probability of brine sand; **f** the probability of oil sand

the iterative procedure. This ensures that non-feasible transitions (e.g., water-saturated rocks on top of hydrocarbon-saturated rocks) do not occur. One advantage of the proposed method is the ability to perturb facies and rock properties and with a single objective function based on the similarity between real and synthetic seismic data. This ensures that both domains are updated and perturbed consistently.

8.3 A Practical Guide for Facies Prediction

While the practical recommendations described in Sect. 7.6 for seismic inversion hold, when predicting facies from seismic data there are additional challenges that need to be considered.

First, the facies one aims to predict from the seismic data should be distinguishable and exhibit a small overlap in the elastic domain. If they have a large overlap they would hardly be predicted individually from the seismic data. Most of the times this step results in aggregating sedimentary facies from the well-log suite into larger facies that are relevant in terms of

sedimentary and fluid composition (i.e., lithofluid facies).

In these types of inversion methodologies, there is the need to rely on a facies-dependent rock physics model to link the rock and the elastic domains. The rock physics models are calibrated at the well location using the available log set and often involves a trial-and-error approach where the model parameters are fitted manually. Nevertheless, the calibrated rock physics model will always have a misfit between predicted and observed logs (Fig. 8.9). These differences might be important and impact negatively the petrophysical predictions from the seismic inversion. A better workflow would be to incorporate the mismatch in the inversion workflow and propagate the uncertainty related to the calibration throughout the entire inversion workflow.

When facies are classified from rock property volumes and not updated directly from the data mismatch, we need to ensure the joint distribution between each property and the relative proportions of each facies as observed from the well-log data (i.e., the training data set) are reproduced in all the models generated during the inversion.

This is of upmost importance to get geologically consistent facies models and increase the success of the inversion. Many seismic inversion methods do fail to reproduce complex relationships due to underlying assumptions about the distribution of the variables of interest. For example, it is rather common to assume Gaussian (or multi-Gaussian) distributions for the rock property distributions when in fact they are frequently multi-model and skewed.

References

Avseth P, Mukerji T, Mavko G (2005) Quantitative seismic interpretation: applying rock physics tools to reduce interpretation risk. Cambridge University Press, Cambridge

Azevedo L, Grana D, Amaro C (2019) Geostatistical rock physics AVA inversion. Geophys J Int 216(3):1728–1739

Azevedo L, Grana D, Figueiredo L (2020) Stochastic perturbation optimization for discrete-continuous inverse problems. Geophysics 85(5):M73–M83. https://doi.org/10.1190/geo2019-0520.1

Batzle M, Wang Z (1992) Seismic properties of pore fluids. Geophysics 57(11):1396–1408

Bjørke K (2009) Reservoir management on Norne—an introduction to time-lapse seismic and EnKF. Master's thesis, Norwegian University of Science and Technology

Bosch M (2004) The optimization approach to lithological tomography: combining seismic data and petrophysics for porosity prediction. Geophysics 69:1272–1282. https://doi.org/10.1190/1.1801944

Bosch M, Mukerji T, Gonzalez EF (2010) Seismic inversion for reservoir properties combining statistical rock physics and geostatistics: a review. Geophysics 75(5):75A165-75A176

Caers J, Hoffman TB (2006) The probability perturbation method: a new look at Bayesian inverse modeling. Math Geol 38(1):81–100. https://doi.org/10.1007/s11004-005-9005-9

Connolly PA, Hughes MJ (2016) Stochastic inversion by matching to large numbers of pseudo-wells. Geophysics 81(2):M7–M22

Cyz M, Azevedo L (2020) Direct geostatistical seismic amplitude versus angle inversion for shale rock properties. IEEE Trans Geosci Remote Sens 59(6):5335–5344. https://doi.org/10.1109/TGRS.2020.3017091

Dadashpour M (2009) Reservoir characterization using production data and time-lapse seismic data. PhD thesis, Norwegian University of Science and Technology

de Figueiredo LP, Grana D, Roisenberg M, Rodrigues BB (2019) Multimodal Markov chain Monte Carlo method for nonlinear petrophysical seismic inversion. Geophysics 84(5):M1–M13

Deutsch C, Journel AG (1992) GSLIB: geostatistical software library and users' guide. Oxford University Press, New York

Elfeki A, Dekking M (2001) A Markov chain model for subsurface characterization: theory and applications. Math Geol 33:569–589

Fjeldstad T, Omre H (2020) Bayesian inversion of convolved hidden Markov models with applications in reservoir prediction. IEEE Trans Geosci Remote Sens 58(3):1957–1968

González EF, Mukerji T, Mavko G (2008) Seismic inversion combining rock physics and multiple-point geostatistics. Geophysics 73(1):R11–R21. https://doi.org/10.1190/1.2803748

Grana D, Della Rossa E (2010) Probabilistic petrophysical-properties estimation integrating statistical rock physics with seismic inversion. Geophysics 75(3):O21–O37

Grana D, Mukerji T, Dvorkin J, Mavko G (2012) Stochastic inversion of facies from seismic data based on sequential simulations and probability perturbation method. Geophysics 77(4):M53–M72

Grana D, Lang X, Wu W (2017) Statistical facies classification from multiple seismic attributes: comparison between Bayesian classification and expectation-maximization method and application in petrophysical inversion. Geophys Prospect 65(2):544–562

Hastie T, Tibshirani R, Friedman J (2009) The elements of statistical learning: data mining, inference, and prediction. Springer, New York

Horta A, Soares A (2010) Direct sequential Co-simulation with joint probability distributions. Math Geosci 42(3):269–292. https://doi.org/10.1007/s11004-010-9265-x

Kemper M, Gunning J (2014) Joint impedance and facies—seismic inversion redefined. First Break 32(9):89–95

Kjølhamar B, Ramirez AC, Jansen S (2020) Seismic Acquisition and Processing: The Technology Race. GEO Magazine

Krumbein WC, Dacey MF (1969) Markov chains and embedded Markov chains in geology. J Int Assoc Math Geol 1(1):79–96

Lindberg DV, Grana D (2015) Petro-elastic log-facies classification using the expectation–maximization algorithm and hidden Markov models. Math Geosci 47(6):719–752

Mavko G, Mukerji T, Dvorkin J (2009) The rock physics handbook—tools for seismic analysis of porous media, 2nd edn. Cambridge University Press

Nunes R, Soares A, Azevedo L, Pereira P (2017) Geostatistical seismic inversion with direct sequential simulation and co-simulation with multi-local distribution functions. Math Geosci 49(5):583–601

Shuey RT (1985) A simplification of the Zoeppritz equations. Geophysics 50:609–614

Soares A (2001) Direct sequential simulation and cosimulation. Math Geol 33(8):911–926

Suman A, Mukerji T (2013) Sensitivity study of rock-physics parameters for modeling time-lapse seismic response of Norne field. Geophysics 78(6):D511–D523. https://doi.org/10.1190/GEO2013-0045.1

The Way Forward

9

9.1 Full Waveform Inversion

Having the tools to build an accurate and detailed velocity model is valuable for seismic processing, imaging and modelling but it can be difficult to achieve, especially in areas of complex subsurface geology. A velocity model is referred to as an Earth model and is a three-dimensional representation of the velocity variations in the subsurface driven by the geology. Geoscientists know the subsurface can be highly complex and it is therefore often represented in a simplified form due to a lack of knowledge and/or data availability to guide spatial complexity. In Chap. 2. the concept of migration showed that velocities impact where the reflectors are positioned in a final seismic image and these data form the basis for interpretative of geology (stratigraphic and structural) and rock properties (quantitative interpretation). Seismic processing and imaging advances in the last few decades have focused on improving the clarity and resolution of reflection events, but in recent years Full Waveform Inversion (FWI) has become a key focus, with accessibility to products greatly increasing. Once, it was only the focus of universities and specialist research groups looking at small, simple and shallow (up to 500 ms TWT) one-dimensional or two-dimensional problems, but in recent years it has become a mainstream commercial offering during seismic imaging projects due to the generalization of cloud computing and high-performance computing.

The full waveform inversion, tries to model all the parts of the wavefield recorded from a seismic shot for a particular range of frequencies. There is a lot of information within the seismic waveform which is currently not used in standard data processing sequences and inversion algorithms. The benefits and promises of this technology are huge, and the energy industry is just starting to unlock them.

9.1.1 Acoustic FWI Background

Although much of the visible progress in the field of FWI has been made during the mid to late twenty–tens, it is not a new concept and the foundations of the modern FWI algorithms were conceived as far back as 40 years. The work by Lailly (1983) describes a method of using a three-dimensional inversion problem to produce a spatially varying model of a medium that represents propagation velocities. The same concept was also published at around the same time by the seminal work of Tarantola (1984) in which he proposes a method to try the "ambitious" task of creating an Earth model from a simple starting model with more than just primary seismic reflections using inversion. The basic equation from these works is shown below:

$$x(\boldsymbol{m}) = \sum_{s.r}[\boldsymbol{d}(t) - \boldsymbol{v}(t, \boldsymbol{m})]^2, \qquad (9.1)$$

© The Author(s), under exclusive license to Springer Nature Switzerland AG 2022
T. Tylor-Jones and L. Azevedo, *A Practical Guide to Seismic Reservoir Characterization*,
Advances in Oil and Gas Exploration & Production, https://doi.org/10.1007/978-3-030-99854-7_9

Fig. 9.1 (Top) Legacy velocity volume from tomography workflow, (Bottom) Updated velocity model using FWI which is more conformable with geology and reflectivity (Bate and Perz 2021). (Seismic Data Courtesy of TGS)

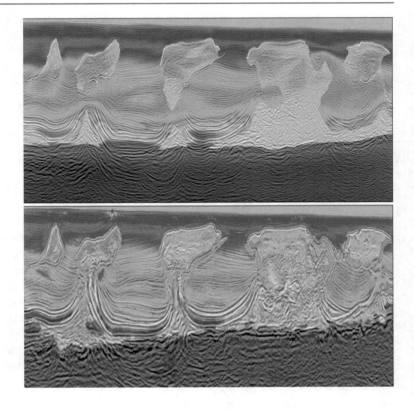

where x is the cost function, d the recorded seismic data, v the synthetic seismic, m the velocity model, t the travel time and s and r the source and receiver, respectively.

Both works describe the application of a global inversion method to iteratively update a velocity model, starting with a simplistic velocity model, using a least-square minimization of the misfit residual difference between the predicted and the observed data. A backpropagation method computes the gradient of the misfit function along which a velocity perturbation model searches to minimize the misfit. The velocity perturbation model is generated after the first iteration, which is added to the initial velocity model to result in an updated model. This updated model is then used as the starting model for the next iteration. At the time, this approach was a very new way of thinking; today

it is clear that the lack of computational power, detailed measurements of seismic survey geometries and suitable seismic data hindered the ability to perform any real data tests.

Further research of the time soon came to similar realizations. It was discovered that when using seismic data with small or medium offsets the inversion process would suffer from getting stuck in local minima within the global inversion space, thus preventing model convergence (Gauthier et al. 1986). By the late 1990s full waveform inversion techniques, supported by advances in computer power, had begun to produce some real two-dimensional data results. A notable example comes from Bunks et al. (1995) which, although it uses a synthetic data set derived from a pre-existing velocity model, shows an insight into the capability for the technique to produce complex, highly detailed velocity models. Pratt (1999) produced several works including one showing an example of a two-dimensional comparison between a tomography velocity model and FWI velocity model for a borehole VSP. Pratt reviewed the current state of FWI technology and included factors such as advances in finite-difference modelling, cheaper computer random access memory, new matrix-based algebra and better artefact minimization as key advances of the success of FWI. It was observed that the issue of getting stuck in local minimum was less of a problem when solving for lower frequencies.

Wide-aperture seismic data with its longer offset was richer in low-frequencies and could also provide both transmitted or refracted/diving waves, which helped the FWI to accurately solve at much lower frequencies. The availability of data containing more reliable low-frequencies (1–4 Hz) led to the adoption of a what is termed a multi-scale approach. Low-frequencies have a far reduced risk of issues of local minima as they are described in the inversion space to have wide valleys in the objective function space. The multi-scale approach accurately solves for the very low-frequencies first and then moves on to solve for the higher frequencies (Bunks et al. 1995).

One of the first implementations of FWI in three-dimensional data was presented in the work by Ben Hadj Ali et al. (2007), where they discuss a method of three-dimensional acoustic finite differences in the wave propagation modelling using the frequency domain method. Although this test used a highly paralleled software solution and ran on a high-performance supercomputer, performance issues with computer memory limitation were raised, which highlights just how complex and computer intensive FWI is. This persistent performance issue is also commented upon in several other works around this time, citing the large computational cost of the forward problem. Specifically, the problem is posed by the intensive calculations required for large number of sources for every iteration of seismic full waveform modelling (Ben Hadj Ali et al. 2009). Other contemporaneous three-dimensional application examples are discussed

in publications by Sirgue et al. (2008) and Warner et al. (2008).

In the more recent times, the focus for acoustic FWI has been to add complexity to explain the wavefield through ever more complete numerical-modelling solutions using finite-difference and finite-element approaches. Additional data types such as early arrivals, diving waves, multiples and refractions along with the reflections are now being included, as they all contain information about the subsurface. Advances in techniques for inversion stability around the issue of cycle skipping have also been a focus, which now help stability issues for low-frequencies to be solved with a very simple starting model. Low-frequency FWI models can now be generated in the very early stages of processing to support velocity model building (Vigh et al. 2019). The continued progress and industry interest in FWI influence seismic acquisition methods which need to be optimized to produce the long offset and low-frequency data required to achieve good FWI results. Current research areas focus on very high-frequency (60–80 Hz) results coming out of FWI.

9.1.2 Velocity Tomography

Seismic reflection-based tomography has been used for many years in seismic processing to produce three-dimensional Earth models that contain a spatial distribution of velocity information. The method uses ray tracing to model the recorded wavefield and can use seismic reflections and refractions to construct the velocity model. Similar to FWI (Sect. 9.1.1), the tomography technique solves an inverse problem by comparing the fit of the predicted data computed from an initial Earth model to the recorded data and then minimizes a misfit function through several iterations.

Velocity tomography is a widely used technique that incorporates additional complexity such as Q-compensation and anisotropy. The velocity tomography workflow is relatively easy to parameterize and run. Tomography produces a superior level of velocity detail and variability

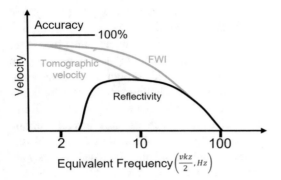

Fig. 9.2 The accuracy and resolution range of tomography, migration velocities (Adapted from Claerbout 1985; Biondi and Almomin 2013)

compared to alternatives coming from interpolations of well-logs or NMO correction velocities, all of which make it the current workhorse of most modern seismic velocity model builds. The technique can however struggle when signal-to-noise is poor and in areas of strong attenuation. By solving the wave equation in a more complete way, FWI goes way beyond the level of velocity detail provided by the ray-tracing approximation of the wave equation used in tomography (Fig. 9.1). FWI can also be taken to much higher frequencies (similar to seismic reflectivity) than tomography is able to reach (Fig. 9.2).

9.1.3   Acoustic FWI Basic Workflow

The basic workflow for FWI is shown in Fig. 9.3. The main input data into the FWI workflow is shot gathers along with a starting velocity model, source wavelet and the position of sources and receivers. Because FWI does not require fully processed seismic data as an input the workflow can be started very early in the seismic processing flow from a fast-track seismic volume. Such a volume can be produced in a quarter of the time than that of the final migrated product and it may give valuable early insight into the subsurface.

Creating the intial model is one of the most critical steps in the FWI process. The starting model is a very simplified velocity model and is often derived from a tomography velocity model that has been heavily smoothed. Other velocity fields such as smooth stacking-velocities or even a smooth well-based model can also be used. Using a multi-scale approach, we fit the lowest frequencies (2–4 Hz) in the data first. Analysis of the seismic data is carried out during this phase of the workflow to understand the most stable low-frequency to model with the first iteration.

The forward model simulates a shot starting with the direct arrival, diving wave and water bottom multiple that uses a numerical solution of the acoustic wave equation. Real shot data and modelled shot data are then compared and the residuals calculated. The residuals are what drives FWI updates because the predicted synthetics should match the real data. To create the update to the initial model the residuals are used to minimize the objective function by perturbating the velocity, which is expressed in the updated model. To ensure the model updates are moving in the right direction a gradient function is calculated between models for each iteration, ensuring that the inversion moves, hopefully, towards a global minimum solution. This process is iterated until the residual stops decreasing or is reduced to an acceptable threshold and the model is converged. So far, FWI is deterministic.

Cycle skipping is due to the oscillatory nature of the seismic data and should be avoided especially at the lower frequencies. Cycle skipping occurs when the phase of the real data and modelled data are out by more than half a wavelength. This is enough of a mismatch to cause inaccurate future model updates, which could ultimately lead to the model converging to the wrong solution (e.g., local minimum) and to inaccuracies in the imaging of the seismic if the velocities are used in the migration. The problem can be approached by trying to generate an initial model with enough accuracy to produce a prediction within half a cycle from the recorded data. This is very challenging and time consuming. A more widely used method is to start with the very lowest usable frequencies as defined by the multi-scale approach. The half cycles are wider at lower frequencies and, therefore, a gradual increase in frequency for

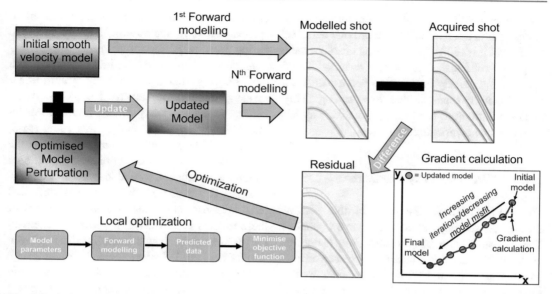

Fig. 9.3 A simplified work of an acoustic FWI workflow

progressive iterations reduces the risk of cycle skipping, though a point will ultimately come when it becomes unavoidable (Yao et al. 2018).

Amplitude discrepancy can also be a common problem in FWI. This occurs when the synthetic data and the real data have different magnitudes of amplitude, commonly when there are sharp contrasts/boundaries in the subsurface or isolated geo-bodies, such as salt. This may also lead to an inaccuracy in the velocity model. To overcome this problem the boundaries of geo-bodies are usually manually added, guided by seismic reflectivity interpretation, to a previously identified stable model update prior to the amplitude issue. This can be time consuming and often requires testing to get the right balance between detail and smoothness to allow the FWI to continue in the right direction.

9.1.4   Acoustic FWI Applications

The integrated application of FWI is a current focus of the energy industry. The technology is a perfect accompaniment to the simultaneous advances happening in seismic acquisition, specifically in the offshore setting, such as broadband data, OBC and OBN data sets

(Chap. 3) with their long offset and ultra-low frequencies. There are several areas where FWI is being applied: Seismic migration (RTM, LSQ); reservoir characterization; shallow hazards and seafloor mapping; pore-pressure prediction; and pseudo-reflectivity.

Technology limits do, however, persist for FWI with the biggest being the computational cost, which exponentially increases with every additional one hertz increase in frequency content that the FWI must solve. The large computer cost creates a high price for producing volumes even at fairly modest frequencies (e.g., 15–20 Hz), making their value questionable against cheaper alternatives such as tomography for use in seismic migration. This cost also restricts the size of data that can be taken to high frequencies using FWI. Seismic data suitability is also another key constraint of FWI. The input data must have long offsets and good deep penetrating diving waves to allow the best stable converged FWI results to be obtained. These properties may be present in modern data sets but the lack of these essential elements in older surveys prohibits the FWI model being pushed much beyond frequencies of modern tomography limits. The depth of an FWI model update is, as a rule of thumb, 1/3 the maximum survey cable length, so for deeper geology it may not be suitable. For interpretation the FWI result can provide detail of layers, but it is difficult to use as an absolute rock property volume as it is a combination of velocity and density information that are difficult to completely separate due to coupling. This makes it difficult to use in a quantitative way for reservoir characterization purposes.

9.1.5   FWI: The Way Forward

The FWI workflow discussed in this section focuses on solving the acoustic wave equation. The acoustic waves offer information relating to velocity and density contrast in the subsurface which cause the reflection events seen in the seismic. Including terms in the objective function to deal with anisotropy and Q-attenuation, as well as using geological constants and models for density estimation further increase the accuracy of an acoustic FWI model. However, the true power of FWI comes when considering elastic FWI to solve the elastic wave equation, which produces a three-dimensional model of the elastic properties of the rocks and pore fluids. The model shows the true rock properties for P- and S-waves, thus, allowing definitive quantitative reservoir characterization to be performed in the subsurface. This takes into account not only all of the P-waves but also all the converted PS-waves (Chap. 2) and so becomes incredibly complex to model when you include waves such as PPS-wave splitting along with all the possible ray paths for reflections, refractions, wave dispersion, wave scattering, absorption effects, etc. Current progress of elastic FWI is in the realm of small three-dimensional models. But with continual advances in quantum computing, multiphysics simulations and finite-element modelling this is getting closer to becoming ever more achievable.

9.2   Handling Uncertainties

Modelling subsurface systems such as those associated with hydrocarbon accumulations, CO_2 reservoirs and aquifers are uncertain. The uncertainty associated with the subsurface models originated in measurement errors recorded during data acquisition, numerical approximations during data processing and modelling, and the lack of knowledge about the system under investigation (Caers 2011; Tompkins et al. 2011). This uncertainty is not homogeneously distributed along the entire modelling grid, instead it is larger for those areas far from the direct observations (i.e., borehole locations) where the number of available conditioning data is small (Caers 2011). While a consistent and reliable assessment of the intrinsic uncertainty of these models is non-trivial it somehow needs to be performed to better quantify risks (Doyen 2007; Caers 2011; Demyanov and Arnold 2018).

Despite limitations regarding uncertainty assessment, modern three-dimensional numerical models are a hub for geoscientists and a mandatory tool for decision-making (e.g., Pyrcz and Deutsch 2002). These models are extremely valuable tools for a reliable description of the spatial distribution of the subsurface rock properties distribution such as porosity, facies and fluids saturation. Depending on data availability, and the assumptions one makes, a geoscientist might use alternative modelling tools with different degrees of complexity. Models built exclusively based on information retrieved from sparsely located direct observations such as wells, and possibly conditioned to a secondary variable of interest (e.g., a trend map or geological model) are always a smooth representation of a much more complex subsurface, and might have limited usability in challenging geological settings and deep targets.

Geophysical data do have the potential to decrease uncertainty predictions when integrated during the geo-modelling workflow. Despite the indirect nature of these data, they cover larger spatial areas when compared with the borehole data and, therefore, might help to constrain the modelling solutions at these locations. However, combining well and geophysical data is not straightforward. There are differences in scale and resolution, and both data are affected by different types of uncertainty levels (Chap. 6). Also, predictions about the subsurface properties from geophysical data require solving a challenging inverse problem (Sect. 9.2).

Stochastic modelling tools, including stochastic geophysical inversion, have the potential to integrate both sets of data, ensure the propagation of the uncertainty throughout the modelling workflow and predict multiple scenarios (i.e., an ensemble of numerical three-dimensional models) that fit equally well the data. To assess uncertainties, this ensemble could be explored in detail using multidimensional data analytics techniques (e.g., Caers 2011; Azevedo et al. 2014; Demyanov and Arnold 2018), such as multidimensional scaling (e.g., Suzuki and Caers 2008; Azevedo et al. 2014; Scheidt et al. 2015) and self-organizing maps (e.g., Phelps et al. 2018). Park et al. (2016) provide a MATLAB® toolbox with spatial data analytics methods that can be used to this end.

However, there are additional layers of uncertainty that are often overlooked in subsurface modelling; particularly those associated with decisions made during the data processing. In seismic reflection data processing a critical step is velocity analysis. The resulting velocity model does impact the quality of the final image obtained after applying imaging algorithms such as migration. The location, shape and amplitude content of the seismic events, which are used in data interpretation and modelling, are highly affected by the velocity. However, its estimation is uncertain and highly dependent on the experience of the seismic processor. Few works have been published to address this challenge (e.g., Li et al. 2014, 2015).

Additionally, seismic interpretation is another step of the geo-modelling that is highly subjective and dependent on decisions made by the seismic interpreter. Bond et al. (2007) perfectly illustrate the conceptual uncertainty in seismic interpretation. How we can include this uncertainty in the geo-modelling workflow and propagate it into geophysical inversion is still an open, and often ignored, question.

As a takeaway message the age of single, smooth and low-spatial resolution is gone. As we move into deeper and more complex geological settings assessing uncertainties is mandatory. In this respect, deterministic modelling tools are no longer suitable and stochastic ones, as provided by geostatistical tools (Deutsch and Journel 1992, Goovaerts 1997; Azevedo and Soares 2017; Grana et al. 2021), should be preferable.

9.3 Machine Learning and Spatial Data Analytics

While this book focuses on conventional methods to integrate seismic reflection data into the geo-modelling workflow, it is impossible to ignore the rapid transformation we are assisting in geophysics with the application of artificial intelligence and spatial data analytics algorithms (e.g., Li et al. 2020). These new techniques leverage recent developments in deep learning neural networks, the computational resources available today in academia and industry and the big data nature of seismic data.

Currently, deep learning methods cover the entire value chain of seismic reflection data: from processing-to-interpretation-to-inversion. Due to the rapid development and the number of new daily publications and pre-prints about new methods that use machine learning in geosciences, we do not aim to provide and extensive overview of the existing methods (we would not be able to do it) but to refer to some of today's most important research avenues that combine machine learning and spatial data analytics with geophysics.

In seismic pre-processing we highlight seismic data interpolation, or regularization. This field has been one of the preferred areas of application of deep learning methods in geophysics. The challenging task of filling-in gaps within the seismic data have been tackled with deep convolutional neural networks (e.g., Wang et al. 2019; Zhang et al. 2020), generative adversarial networks (e.g., Oliveira et al. 2018; Siahkoohi et al. 2018; Kaur et al. 2021), geostatistical simulation (Turco et al. 2019), support vector regression (e.g Jia and Ma 2017; Jia et al. 2018) and relevance vector machine (e.g., Pilikos 2020). Often, these methods are accompanied with dimensionality reduction techniques and transfer learning methods (e.g., Pan and Yang 2010). Contrary to model-driven or interpolation techniques based on wave-equation-based methods, machine learning algorithms are cheaper to apply after training, are data driven and do not require mathematical transformations into another physical domain (e.g., Fourier, Radon) (e.g., Spitz 1991) or any velocity model.

In seismic interpretation, deep learning methods have been applied to the identification of seismic facies and automatic horizon tracking (e.g., Zhao et al. 2015; Shi et al. 2020; Grana et al. 2020), structural features (e.g., Huang et al. 2017; Di et al. 2018, 2020; Zhao and Mukhopadhyay 2018) and to generate realistic geological models from the data (e.g., Wu et al. 2020). Deep autoencoders and deep convolutional neural networks have been the most promising techniques to accomplish these tasks. Automatic well-log imputation, prediction and classification is another realm where deep learning methods are being widely explored. The following list of references, although not exhaustive, comprises some examples of these methods: Baldwin et al. 1990; Silva et al. 2015; Xie et al. 2018; Dell'Aversana 2019; Chen et al. 2020; Feng et al. 2021.

An overview of the deep learning methods for seismic inversion is provided by Adler et al. (2021). In this book we decided to limit the referencing of deep learning methods for reservoir geophysics (i.e., seismic inversion) to those based on "physics-guided" (or physics-aware) approaches. In this type of approach, the predictions about the subsurface include the integration of a physical model that mimics the system under investigation within the machine learning algorithm and therefore surpasses the need of labelled training data. This set of new tools uses end-to-end neural network architectures (e.g., Biswas et al. 2019; Das et al. 2019). Physics-guided machine learning algorithms are indeed the class of machine learning methods that has the potential to deeply transform geosciences in areas ranging from applied and exploration geophysics to Earth observations (Tuia et al. 2021).

Despite the promises of deep learning and spatial data analytics techniques in subsurface reservoir characterization, there are challenges that still need to be addressed. Often deep neural networks face the problem of overfitting the predictions and have problems in generalization. The selection of the network architecture and hyperparameter tuning might be challenging. Lastly, the integration of uncertainty in the predictions obtained by this type of system is still shallow as well as making these networks interpretable (Toms et al. 2020). However, making robust predictions about the subsurface is critical in geosciences problems for better decision-making (Sect. 1.2).

9.4   Towards Carbon Neutrality

As we continue to understand and try to predict the future energy needs of the planet, we are battling an ever-increasing demand for energy. However, the challenge today is now not only about the quantity of energy required but also being selective as to the impact that producing a particular energy has on the environment. It is commonly believed that moving to renewable sources (i.e., hydro, solar, wind) is the answer, yet as we start to build and use these power sources at ever bigger scales it is becoming evident that they create environmental impacts and require large amounts of mineral resources that must be managed. In the last decade of the twenty-first century the role of the geoscientist within this journey has been brought into focus more than at any other time.

Along with energy production, industrial and manufacturing processes are also under pressure to evolve to meet ever growing environmental concerns. Decarbonisation of industry is a process that recognizes the need for certain activities to exist in the modern world but that these activities are particularly large producers of carbon dioxide. So, by reducing or completely removing their carbon-based emissions these processes become cleaner and transform into carbon neutral processes.

While this book focuses on the description of geo-modelling workflows which have been typically applied in the energy industry, mainly for hydrocarbon reservoir modelling and characterization, this is not the end of line for seismic reflection data and its quantitative interpretation. Most of the techniques and best practices introduced in this book will play a central role in facing the current and future challenges related with the energy transition towards carbon neutrality.

9.4.1 CCUS

Carbon capture, utilization and storage (CCUS) is a process that aims to prevent carbon dioxide from being emitted into the atmosphere, where it may contribute along with other gases such as methane and nitrous oxide to the commonly acknowledged greenhouse effect. The process aims to capture the gas and then transport it for long-term (hundreds of years) storage or to be used for other activities thereby managing it. CCUS plays an important role in the ongoing global efforts to reduce global temperature rises caused by humankind's activities. Geological CO_2 storage is today a real alternative for companies with high-emission rates. To answer these needs, national authorities are building their own inventory of potential injection sites, similar to what happens in hydrocarbon exploration.

The research, investment and application of carbon dioxide storage have greatly increased in the last 5–10 years. Geological storage of CO_2 is already a reality in some regions of the globe with the successful development of commercial pilot sites such as Net Zero Teeside in the UK. Carbon dioxide has historically been used in the energy industry for enhancing the recovery of hydrocarbons from subsurface reservoirs during the production phase. The Sleipner field (e.g., Chadwick et al. 2006; Furre et al. 2017) is the first commercial offshore carbon capture and storage project located in the Norwegian section of the North Sea. This pilot case study has been a flagship project worldwide. The timescale for storing CO_2 is much longer than the life of a hydrocarbon field. While a hydrocarbon field

may be active for 20–40 years, the timescale for CO_2 storage is centuries. This means we need to identify geological stores with the appropriate injectivity and capacity to keep it locked away. Figure 9.4 shows an illustration of how carbon capture can work using a powerplant and industrial processes as the CO_2 source.

Currently there are a few different subsurface storage solutions for CO_2. An anticline structural feature of sufficient size in the subsurface can be used for injecting CO_2 via a suitable geological unit (e.g., saline aquifer). The anticline provides a structural closure mechanism for the CO_2. Another concept is to inject CO_2 into a permeable layer which is bounded by highly impermeable and non-transmissible lithology of appropriate thickness, thereby relying on containment by stratigraphic mechanisms. A combination of both mechanisms is also viable.

A consideration of these methods is how CO_2 injection affects the geology from a geomechanical perspective. If subsurface storage is a viable option for containing CO_2 the process of injection will cause a large increase in formation pressure over time. This impact needs to be fully understood as the possibility of mechanical failure of storage lithologies may lead to the sudden release of CO_2 into the atmosphere.

Challenges also come from the mobility of the CO_2 in the subsurface. CO_2 remains mobile long after injection and being able to adequately model future behaviour over the life of the storage is a challenging exercise. CO_2 is also a reactive substance meaning it can react with the mineralogy of the geology, not to mention that it is highly soluble in brine water.

From a reservoir characterization perspective understanding and characterizing the depositional environment and the rock properties in the subsurface is crucial for determining if an area is suitable for storage. These areas often do not have any borehole penetrations and little is known about the spatial distribution of the subsurface properties. If borehole data do exist, then this is where probabilistic modelling can be an important tool. Seismic data are the main decision tool throughout a CCUS project: from its early stage for defining the sweet spot for carbon

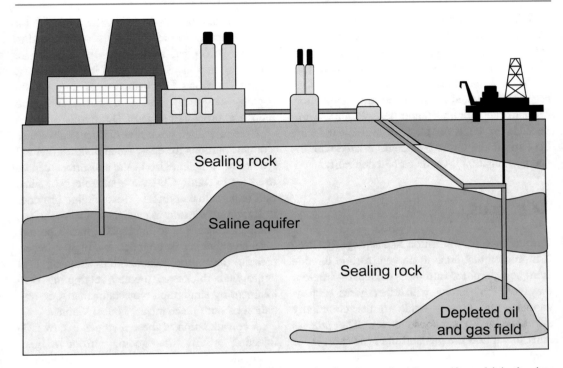

Fig. 9.4 Conceptual diagram illustrating a model for CCUS showing storage in saline aquifer and injection into depleted oil and gas field

dioxide injection and during injection (i.e., geological formations with suitable porosity and permeability), to modelling and forecasting the evolution of the CO_2 plume over time (Furre et al. 2017). The analysis and understanding of faults' location and offset distance is also very important to evaluate the faults seal potential, if required.

9.4.2 Geothermal

The role of reservoir characterization within geothermal energy is something that has been around for a long time. The general concept for geothermal is simple. Geothermal systems involve drilling a minimum of two boreholes (although multiple lateral borehole designs now exist) several kilometres into the subsurface targeting a formation with high temperature above 100 °C. Water is then pumped between the boreholes where it heats up as it travels through the surrounding rocks but remains in a liquid

state due to pressures. As the heated water returns to the surface it turns into high-temperature steam, which is used to drive electrical turbines to produce energy. Any residual water or condensed steam is then re-injected back into the ground to create an open loop system. Figure 9.5 shows the concept of geothermal power generation. Figure 9.6 shows a geothermal power station in Iceland.

Changes in subsurface temperature are caused by geothermal gradient. The geothermal gradient varies around the world but can be seen in ranges of 10–30 °C per kilometre. In very volcanically active places (e.g., Iceland) ground temperature can be much hotter close to the surface. Having the presence of igneous intrusions (e.g., granite) can also lead to high temperatures due to the decay of radioactive elements in the minerals.

Geothermal energy systems can be divided into what is known as conventional and unconventional systems. The conventional systems require no additional intervention to be viable geothermal energy sources, however these

Fig. 9.5 Conceptual diagram
of geothermal power plant

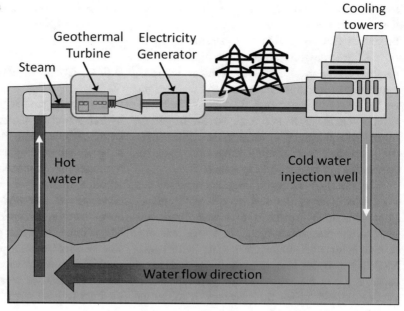

systems are often difficult to find. Unconventional systems, also known as enhanced or engineered geothermal systems (EGS), exploit a wider range of geothermal sources which require additional stimulation to be viable for energy generation, much like in unconventional hydrocarbons. Boreholes are used at much deeper

depths than conventional sources (5–6 km) and, as such, benefit from the much higher temperatures. The concept of EGS was initiated in Los Alamos (USA) and Cornwall (UK) and focuses on drilling two (or more) deep boreholes into lower permeability rock which then requires stimulation by hydraulic fracturing or chemical

Fig. 9.6 The Krafla geothermal power station in north-east Iceland which opened in 1978 and uses steam from 18 borehole to drive two 30 MW turbine (Whaley 2016). (Image courtesy of GEO ExPro)

stimulation. The stimulation in the rocks allows the injected water to flow easily between the two boreholes (Procesi 2014). An extreme example of EGS includes investigating the harnessing of geothermal energy from supercritical extreme temperature systems associated with magma bodies (e.g., hydrothermal fluids) (Wilfred and Frioleifsson 2010; Natale et al. 2013). Targeting highly overpressured systems (Chap. 6) within the subsurface is also one of the more extreme targets for geothermal energy sources.

The role of reservoir characterization for geothermal energy has several different applications. The main aspect is to locate a suitable geothermal reservoir target. Most locations are several kilometres under the surface with some ESG targets being even deeper. Techniques such as gravity and magnetics do not generally have enough resolution for the accuracy required to design borehole placements and may struggle with penetration to image deeper zones (5–6 km) in the subsurface (Barbier 2002). If boreholes already exist, then it is possible to carry out a micro-seismic survey to identify the localized geology in some detail. Passive seismic techniques (Chap. 3) have been applied successfully to help identify suitable geothermal zones within an area of interest (Obermann et al. 2015; Li et al. 2017). Seismic reflection data are an ideal data set for geothermal projects. It can provide the resolution required and allows important interpretation and delineation of things like fracture and fault orientations, which are critical to understand the fluid flow within the rocks.

Seismic data are routinely used for helping to plan and drill successful and safe wells for hydrocarbon development. The identification of drilling hazards and avoidance of large, faulted zones is also a requirement for geothermal borehole drilling.

Seismic inversion techniques (Chap. 7) are also useful for geothermal projects to help understand porosity distributions and identify boundaries and barriers to fluid flow. ESG targets that require mechanical fracture stimulation, like onshore unconventional hydrocarbon reservoirs, will benefit from inversion derived volumes for

properties such as Young's modulus, Possion's ratio and brittleness.

9.4.3 Green Hydrogen

Green hydrogen is hydrogen generated by hydrolyses using the excess of energy produced by renewables, and this is also a field where seismic data might play an important role. Green hydrogen is produced and stored until it is required in the case when energy output from conventional renewable sources is insufficient to cover energy demands. The solution might be below our feet. As in geological carbon dioxide storage (Sect. 9.4.1), porous and permeable formation within the subsurface might be the preferred location to store green hydrogen in the medium to long term.

9.4.4 Wind Energy

The concept of harnessing energy from the wind goes back to first century AD with the invention of the windwheel by Heron of Alexandria. The first, true commercial offshore wind farm was the Vindeby (offshore Denmark) in 1991 which consisted of 11 turbines generating 5 MW of power. Early offshore wind farms were simply onshore turbines on concrete foundations located in relatively shallow water. Today, wind power is an attractive energy investment due to it being both a renewable and zero-carbon energy production source. The discussion around the carbon footprint of turbine manufacturing, maintenance and disposal is something to be aware of and is the topic of numerous current studies (Bhandari et al. 2020; Morini et al. 2021).

Modern windfarm projects, especially offshore, are now getting spatially more vast and include ever larger turbines in deeper water. Examples include the Hornsea wind farm in the UK with a footprint of ~ 1725 km^2. Understanding the seafloor conditions as well as the shallow geology is critical even in the early feasibility stage of selecting a suitable site.

Complexities such as UXO (unexploded ordnance) detection and mitigation need to be carefully managed. This is done with a combination of high-definition geophysical techniques and borehole sampling. The data are then integrated, much like in the hydrocarbon industry, to produce a three-dimensional ground model with information about soil and geotechnical properties. In response to the development of large wind farm projects emerging technologies such as automated surveys for seafloor mapping are now used.

Recent experiences of planning new large-scale wind farm developments have highlighted that variations in ground conditions mean a variety of ground engineering solutions will be required for individual turbines within the same site. Seismic reflection data are a central piece of information for decision-making during the installation of the infrastructure related to this type of renewable energy. In this framework, the focus is on the development of efficient and cheap acquisition schemes of ultra-high-resolution seismic reflection data (i.e., high-frequency data). These data represent a new reality for seismic interpreters and geoscientists, as they have considerable challenges related to the acquisition, processing and prediction of geotechnical properties of the first hundreds of metres below the seafloor.

References

Adler A, Araya-Polo M, Poggio T (2021) Deep learning for seismic inverse problems: toward the acceleration of geophysical analysis workflows. IEEE Signal Process Mag 38(2):89–119. https://doi.org/10.1109/MSP.2020.3037429

Azevedo L, Soares A (2017) Geostatistical methods for reservoir geophysics. Springer

Azevedo L, Nunes R, Correia P, Soares A, Guerreiro L, Neto GS (2014) Multidimensional scaling for the evaluation of a geostatistical seismic elastic inversion methodology. Geophysics 79(1):M1–M10. https://doi.org/10.1190/geo2013-0037.1

Baldwin JL, Bateman RM, Wheatley CL (1990) Application of a neural network to the problem of mineral identification from well logs. Log Anal 31(5):279–293

Barbier E (2002) Geothermal energy technology and current status: an overview. Renew Sustain Energy Rev 6:3–65

Bate D, Perz M (2021) Gulf of Mexico, a Proving Ground for OBN Technology. GEO ExPro 18(2)

Ben Hadj Ali H, Operto S, Virieux J, Sourbier F (2007) 3D acoustic frequency-domain full-waveform inversion. SEG Technical Program Expanded Abstracts, pp 1730–1734

Ben Hadj Ali H, Operto S, Virieux J (2009) Efficient 3D frequency-domain full waveform inversion (FWI) with phase encoding: 71st EAGE annual conference and exhibition (Amsterdam, 8–11 June), Expanded Abstracts, p 5812

Bhandari R, Kumar B, Mayer F (2020) Life cycle greenhouse gas emission from wind farms in reference to turbine sizes and capacity factors. J Clean Prod vol 277:123385

Biondi B, Almomin A (2013) Tomographic full-waveform inversion (TFWI) by combining FWI and wave-equation migration velocity analysis. Lead Edge 32(9):1074–1080

Biswas R, Sen MK, Das V, Mukerji T (2019) Prestack and poststack inversion using a physics-guided convolutional neural network. Interpretation 7(3):SE161–SE174

Bond C, Gibbs A, Shipton Z, Jones S (2007) What do you think this is? "Conceptual Uncertainty" in geoscience interpretation. GSA Today 17(10):4–10. https://doi.org/10.1130/GSAT01711A.1

Bunks C, Saleck FM, Zaleski S, Chavent G (1995) Multiscale seismic waveform inversion. Geophysics 60(5):1457–1473

Caers J (2011) Modeling uncertainty in earth sciences. UK, Wiley-Blackwell

Chadwick A, Arts R, Eiken O, Williamson P, Williams G (2006) Geophysical monitoring of the CO2 plume at Sleipner, North Sea: an outline review. In: Lombardi S, Altunina LK, Beaubien SE (eds) Advances in the geological storage of carbon dioxide: international approaches to reduce anthropogenic greenhouse gas emissions. Springer, Dordrecht, Netherlands, pp 303–314

Chen W, Yang L, Zha B, Zhang M, Chen Y (2020) Deep learning reservoir porosity prediction based on multilayer long short-term memory network. Geophysics 85:WA213–WA225

Claerbout JF (1985) Imaging the earth's interior. Blackwell Scientific Publications

Das V, Pollack A, Wollner U, Mukerji T (2019) Convolutional neural network for seismic impedance inversion. Geophysics 84:R869–R880

Dell'Aversana P (2019) Comparison of different machine learning algorithms for lithofacies cladrassification from well logs. Bollettino di Geofisica Teorica ed Applicata 60(1):69–80. BGTAAE0006-6729

Demyanov V, Arnold D (2018) Challenges and solutions in stochastic reservoir modelling: geostatistics, uncertainty prediction. EAGE Publishing BV, Machine Learning

Deutsch C, Journel AG (1992) GSLIB: geostatistical software library and users' guide. Oxford University Press, New York

Di H, Shafiq M, AlRegib G (2018) Patch-level MLP classification for improved fault detection. In: 88th annual international meeting. SEG Technical Program Expanded Abstracts, pp 2211–2215. https://doi.org/10.1190/segam2018-2996921.1

Di H, Li Z, Maniar H, Abubakar A (2020) Seismic stratigraphy interpretation by deep convolutional neural networks: a semisupervised workflow. Geophysics 85(4):WA77–WA86. https://doi.org/10.1190/geo2019-0433.1

Doyen P (2007) Seismic reservoir characterization: an earth modelling perspective. EAGE

Feng F, Grana D, Balling N (2021) Imputation of missing well log data by random forest and its uncertainty analysis. Comput Geosci 152:104763

Furre A-K, Eiken O, Alnesa H, Vevatnea JN, Kiæra AF (2017) 20 years of monitoring CO2-injection at Sleipner. Energy Procedia 114:3916–3926

Gauthier O, Virieux J, Tarantola A (1986) Two-dimensional nonlinear inversion of seismic waveforms: numerical results. Geophysics 51(7):1387–1403

Goovaerts P (1997) Geostatistics for natural resources evaluation. New York: Oxford University Press.

Grana D, Azevedo L, Liu M (2020) A comparison of deep machine learning and Monte Carlo methods for facies classification from seismic data. Geophysics 85:WA41–WA52

Grana D, Mukerji T, Doyen P (2021) Seismic Reservoir Modeling: Theory, Examples, and Algorithms. Wiley-Blackwell.

Huang L, Dong X, Clee TE (2017) A scalable deep learning platform for identifying geologic features from seismic attributes. Lead Edge 36(3):249–256. https://doi.org/10.1190/tle36030249.1

Jia Y, Ma J (2017) What can machine learning do for seismic dataprocessing? An Interpolation Application. Geophysics 82(3):V163–V177. https://doi.org/10.1190/geo2016-0300.1

Jia Y, Yu S, Ma J (2018) Intelligent interpolation by Monte Carlo machine learning, An interpolation application. Geophysics 83(2):V83–V97. https://doi.org/10.1190/geo2017-0294.1

Kaur H, Pham N, Fomel S (2021) Seismic data interpolation using deep learning with generative adversarial networks. Geophys Prospect 69(2):307–326. https://doi.org/10.1111/1365-2478.13055

Lailly P (1983) The seismic inverse problem as a sequence of before stack migrations. In: Bednar JB, Robinson E, Weglein A (eds) Conference on inverse scattering—theory and application.Society of Industrial and Applied Mathematics, Expanded Abstracts, pp 206–220

Li L, Caers J, Sava P (2014) Uncertainty maps for seismic images through geostatistical model randomization SEG Technical Program Expanded Abstracts, pp 1496–1500

Li L, Caers J, Sava P (2015) Assessing seismic uncertainty via geostatistical velocity-model perturbation and image registration: an application to subsalt

imaging. The Leading Edge 34(9):1064–1066, 1068–1070. https://doi.org/10.1190/tle34091064.1

Li Z, Bao F, Zhang S, Jia X, Yuen D (2017) Seismic imaging for the geothermal resources with dense seismic array and passive sources. SEG Global Meeting Abstracts: 867–870.

Li W, Hu W, Abubakar A (2020) Machine learning and data analytics for geoscience applications—introduction. Geophysics 85:WAi–WAii

Morini AA, Ribeiro MJ, Hotza D (2021) Carbon footprint and embodied energy of a wind turbine blade—a case study. Int J Life Cycle Assess 26:1177–1187. https://doi.org/10.1007/s11367-021-01907-z

Natale G de, Troise C, Troiano A (2013) Campi Flegrei deep drilling project and geothermal activities in Campania Region (Southern Italy). Geophys Res Abstr 15:EGU2013-13155-4

Obermann A, Kraft T, Larose E, Wiemer S (2015) Potential of ambient seismic noise techniques to monitor the St. Gallen geothermal site (Switzerland). JGR Solid Earth 120(6):4301–4316

Oliveira DAB, Ferreira RS, Silva R, Brazil EV (2018) Interpolating seismic data with conditional generative adversarial networks. IEEE Geosci Remote Sens Lett 15(12):1952–1956. https://doi.org/10.1109/LGRS.2018.2866199

Pan SJ, Yang Q (2010) A survey on transfer learning. IEEE Trans Knowl Data Eng 22(10):1345–1359. https://doi.org/10.1109/TKDE.2009.191

Park J, Yang G, Satija A, Scheidt C, Caers J (2016) DGSA: a Matlab toolbox for distance-based generalized sensitivity analysis of geoscientific computer experiments. Comput Geosci 97:15–29. https://doi.org/10.1016/j.cageo.2016.08.021

Phelps G, Scheidt C, Caers J (2018) Exploring viable geologic interpretations of gravity models using distance-based global sensitivity analysis and kernel methods. Geophysics 83(5):G79–G92

Pilikos G (2020) The relevance vector machine for seismic Bayesian compressive sensing. Geophysics 854:WA279–WA292

Pratt, RG (1999) Seismic waveform inversion in the frequency domain, part 1: theory and verification in a physical scale model: Geophysics 64(3):888–901. https://doi.org/10.1190/1.1444597

Procesi M (2014) The unconventional geothermal resources: features and current uses. Energy Science and Technology, National institute of Geophysics and Volcanology, Stadium Press LLC

Pyrcz MJ, Deutsch C (2002) Geostatistical reservoir modeling, 2nd edn. Oxford University Press, New York

Scheidt C, Jeong C, Mukerji T, Caers J (2015) Probabilistic falsification of prior geologic uncertainty with seismic amplitude data: application to a turbidite reservoir case. Geophysics 80(5):M89–M12. https://doi.org/10.1190/geo2015-0084.1

Shi Y, Wu X, Fomel S (2020) Waveform embedding: automatic horizon picking with unsupervised deep learning. Geophysics 85(4):WA67–WA76

Siahkoohi A, Kumar R, Herrmann FJ (2018) Seismic data reconstruction with generative adversarial networks. In: 80th EAGE conference and exhibition 2018, pp 1–5

Silva AA, Lima Neto IA, Misságia R, Ceia MA, Carrasquilla AG, Archilha NL (2015) Artificial neural networks to support petrographic classification of carbonate-siliciclastic rocks using well logs and textural information. J Appl Geophys 117:118–125. https://doi.org/10.1016/j.jappgeo.2015.03.027

Sirgue L, Etgen JT, Albertin U (2008) 3D frequency domain waveform inversion using time domain finite difference methods. In: 70th EAGE conference and exhibition incorporating SPE EUROPEC 2008, Roma, Italy, Expanded Abstracts, p F022

Spitz S (1991) Seismic trace interpolation in the F-X domain. Geophysics 56(6):785–794. https://doi.org/10.1190/1.1443096.GPYSA70016-8033

Suzuki S, Caers J (2008) A distance-based prior model parameterization for constraining solutions of spatial inverse problems. Math Geosci 40(4):445–469. https://doi.org/10.1007/s11004-008-9154-8

Tarantola A (1984) Inversion of seismic reflection data in the acoustic approximation. Geophysics 49(8):1259–1266. https://doi.org/10.1190/1.1441754

Tompkins MJ, Fernández Martínez JL, Alumbaugh DL, Mukerji T (2011) Scalable uncertainty estimation for nonlinear inverse problems using parameter reduction, constraint mapping, and geometric sampling: marine controlled-source electromagnetic examples. Geophysics 76(4):F263–F281

Toms BA, Barnes EA, Ebert-Uphoff I (2020) Physically interpretable neural networks for the geosciences: applications to earth system variability. J Adv Model Earth Syst 12(9):1–20. https://doi.org/10.1029/2019MS002002

Tuia D, Roscher R, Wegner JD, Jacobs N, Zhu X, Camps-Valls G (2021) Toward a collective agenda on AI for earth science data analysis. IEEE Geosci Remote Sens Mag 9(2):88–104

Turco F, Azevedo L, Herold D (2019) Geostatistical interpolation of non-stationary seismic data. Comput Geosci 1–18

Vigh D, Cheng X, Jiao K, Sun D (2019) Keys to robust reflection-based full-waveform inversion. In: Conference proceedings, 81st EAGE conference and exhibition 2019, vol 2019, pp 1–5

Wang B, Zhang N, Lu W, Wang J (2019) Deep-learning-based seismic data interpolation: a preliminary result. Geophysics 84(1):V11–V20. https://doi.org/10.1190/geo2017-0495.1.GPYSA70016-8033

Warner M, Stekl I, Umpleby A (2008) Efficient and effective 3D wavefield tomography. In: 70th EAGE conference and exhibition incorporating SPE EUROPEC 2008, Roma, Italy, Expanded Abstracts

Whaley J (2016) Iceland. Harnessing the Earth. GEO ExPro 13(2)

Wilfred AE, Frioleifsson GO (2010) The science program of the Iceland drilling projects (IDDP): a study of supercritical geothermal resource. In: Proceedings world geothermal congress 2010

Wu X, Geng Z, Shi Y, Pham N, Fomel S, Caumon G (2020) Building realistic structure models to train convolutional neural networks for seismic structural interpretation. Geophysics 85(4):WA27–WA39

Xie Y, Zhu C, Zhou W, Li Z, Liu X, Tu M (2018) Evaluation of machine learning methods for formation lithology identification: a comparison of tuning processes and model performances. J Petrol Sci Eng 160:182–193. https://doi.org/10.1016/j.petrol.2017.10.028

Yao G, da Silva NV, Warner M, Kalinicheva T (2018) Separation of migration and tomography modes of full-waveform inversion in the plane wave domain. J Geophys Res Solid Earth 123:1486–1501. https://doi.org/10.1002/2017JB015207

Zhang H, Yang X, Ma J (2020) Can learning from natural image denoising be used for seismic data interpolation? Geophysics 85(4):WA115–WA136

Zhao T, Mukhopadhyay P (2018) A fault-detection workflow using deep learning and image processing. SEG Technical Program Expanded Abstracts, pp 1966–1970. https://doi.org/10.1190/segam2018-2997005.1

Zhao T, Jayaram V, Roy A, Marfurt KJ (2015) A comparison of classification techniques for seismic facies recognition. Interpretation 3(4):SAE29–SAE58. https://doi.org/10.1190/INT-2015-0044.1

Index

T. Tylor-Jones and L. Azevedo, *A Practical Guide to Seismic Reservoir Characterization*,
Advances in Oil and Gas Exploration & Production, https://doi.org/10.1007/978-3-030-99854-7